The Moon: A Popular Treatise

Garrett P. Serviss

Originally Published by
D. APPLETON AND COMPANY
1907

Contents

PREFACE

The reader familiar with astronomical literature will doubtless remark a certain resemblance between the plan on which this book is written and that of Fontenelle's "Conversations on the Plurality of Worlds," a French classic of the eighteenth century. The author freely acknowledges that it was the recollection of the pleasure which the reading of Fontenelle's book gave him, years ago, that led to the adoption of a somewhat similar plan for this description of the moon. But, except that in both cases the conversational method is employed, no great likeness will be found between what is here presented and the work of the witty Frenchman.

Having been invited by the Messrs. Appleton & Co. to prepare a small volume, to be based on a series of lunar photographs representing the moon as it appears on successive evenings during an entire lunation, the author felt that the work should be made as entertaining as possible. He has, therefore, avoided technicalities, while endeavoring to present all the most essential facts known about our satellite. What he has written is intended for the general reader, who desires to learn the results of the great advances in astronomy without being too much troubled with the scientific methods by whose aid those results have been reached.

This is the first time, as far as the author is aware, that a series of lunar photographs, showing our satellite in its varying aspects from New to Old Moon, has been presented in a book, accompanied with a description of the mountains, plains, volcanoes, and other formations shown in each successive photograph. The reader is enabled to place himself, as it were, in an observatory of the first rank, provided with the most powerful apparatus of the astronomer, and, during an entire month, view the moon in her changing phases.

The photographs here reproduced were made at the Yerkes Observatory, and the most grateful acknowledgments are tendered to Prof. Edwin B. Frost, its director, for generously consenting to their use for this purpose. He could only have been induced to do so by his desire to see the fruits of the admirable work accomplished by his associates enjoyed by an ever-widening circle.

The series of photographs representing the moon on successive evenings were taken with the 12-inch telescope of the Yerkes Observatory by Mr. James

1

Wallace, who employed a color filter that he constructed specially for this telescope, which possesses a visual and not a photographic objective. The larger scale photographs, representing certain selected regions on the moon, were taken by Mr. Ritchey, now of the Carnegie Solar Observatory at Mount Wilson, California, with the great 40-inch telescope of the Yerkes Observatory. It is unnecessary to speak of the extraordinary quality of these photographs, which have been admired by astronomers in all lands.

It should, perhaps, be added that while the director of the Yerkes Observatory has shown confidence in the author by intrusting to him the use of these photographs, yet, neither Professor Frost, nor Messrs. Wallace and Ritchey are in any way responsible for the statements made in this book. The author has taken pains to be accurate, but if any errors of fact or opinion have crept in, he alone must be blamed for them.

Garrett P. Serviss.

Château d'Arceau,

Côte d'Or, France, June, 1907.

FULL-PAGE ILLUSTRATIONS Copernicus and the Carpathian Mountains Frontispiece PHOTOGRAPHS OF THE MOON SHOWING PHASES OF CHANGE

INTRODUCTION

INTRODUCTION

ONE serene evening, when the full moon, rising slowly above the tree tops, began to spread over the landscape that peculiar radiance which, by half revealing and half concealing, by softening all outlines, and by imparting a certain mystery to the most familiar objects, fascinates at once the eye and the imagination, I was walking with a friend, a lady of charming intelligence, in a private park adjoining an old mansion in one of the most beautiful districts of central New York. For a long time we both remained silent, admiring the scene before us, so different in every aspect from its appearance in the glare of daylight—each occupied with the thoughts that such a spectacle suggests. Suddenly my friend turned to me and said:

"Tell me—for, like so many thousand others, I am virtually ignorant of these mysteries of the sky—tell me, what is that moon? What do astronomers really know about it?"

"But," I replied, "you certainly exaggerate your ignorance. You must have read what so many books have told about the moon."

"Not a word," was the reply, "or at least, what I have read has made little impression upon my mind. I read few books of science; generally they repel me. But face to face with that marvelous moon, I find it irresistible, and my desire for knowledge concerning it becomes intense. I remember something about eclipses, and something about tides, with which, I believe, the moon is concerned. I recall the statement that the moon has no atmosphere, but does possess great mountains and volcanoes. Yet these things are so jumbled in my memory with technical statements which failed to interest me, that really my ignorance remains profound. But I have heard that many surprising discoveries have been made lately concerning the moon, and that astronomers have succeeded in taking wonderful photographs of scenes in the lunar world. I have, indeed, seen copies of some of these photographs, but beyond awaking curiosity by their *bizarre* effects of light and shadow, they impressed me little, for lack, I suppose, of information as to their meaning. I beg you, then, to tell me what is really known about the world of the moon. There it is; I see it; I experience the delightful impressions which its light produces—but, after all, what is it, and what should we behold if we could go there? I once read Jules Verne's romance of a trip to the moon, but unfortunately his adventurers never really got there, and I finished the story with a keen sense of disappointment because, in the end, he told so very little about the moon itself. As for the professional books of the astronomers they are useless to me. Then, please tell me that which, at this moment, with that wonderful orb actually in sight, I so much desire to know."

It was not possible to resist an appeal so earnestly urged, but I felt compelled to say: "Since you remember so little about the fundamental facts which generations of astronomers have accumulated concerning our nearest neighbor in the sky, I must, for the sake of completeness, and in order to put you *au courant* with the more captivating things that will come later, begin at the beginning, and the true beginning is not among the mountains of the moon, but here on the earth. We must start from our own globe—as the moon herself did."

"What do you mean by that?" my friend asked with a tone of surprise.

"Have you not read, somewhere, in the last ten years, that the moon was actually born from the earth?"

"Yes, now that you mention it, I dimly recall something of the kind, but I took it for an extravagant speculation of some *savant* who possessed more imagination than solid knowledge."

"The *savant* who originally demonstrated the earthly origin of the moon," I replied, "is not one to be easily led into extravagance by his imagination. It is Prof. George Darwin, the son of the famous author of the 'Origin of Species.' I shall not mention his mathematics, which are troublesome, but allow me to tell you, in a word, that his investigations have satisfied astronomers that the earth and the moon once composed a single body. How many million years ago that was we can only guess. The causes of the separation which eventually occurred were the plastic condition of the original body while it was yet hot and molten, its swift axial rotation producing an immense centrifugal force at its equator, and the attraction of the sun raising huge tides which affected its entire mass instead of affecting only the waters of the ocean as the tides do at present. At last there came a time when an enormous portion of the swiftly rotating globe was torn loose. That portion included about one-eightieth of the entire mass of the earth. Some astronomers and geologists think that the 'wound' left in the side of the earth by this stupendous excision is yet traceable in the basin of the Pacific Ocean.

"The separation being once effected, the material that had escaped gradually assumed a globular form under the influence of the gravitation of its own particles; and, at the same time, by virtue of a curious reaction of the tidal attractions of the two bodies upon each other, the new-born globe was slowly forced away from its mother earth, becoming, in fact, its satellite. Thus, by a process which certainly does seem extravagantly imaginative, but which, nevertheless, is approved by strict mathematical deductions from known physical facts, the moon is believed to have had her birth."

"Surely," said my companion, "my imagination would never have dared to form such a picture, even if it had been capable of so extraordinary a flight."

"No," I replied, "nor the imagination of the most learned astronomer. You perceive that in things celestial as in things terrestrial fact is far more strange than fiction. We shall have occasion to refer to some of the consequences of the earthly origin of the moon later on, but just now in order that the knowledge you seek may not be too fragmentary, I must tell you some other, more commonly known, facts about our satellite."

"Judging by myself I doubt if there are many such facts *commonly* known."

"Perhaps you are right, but do not judge too severely the authors of astronomical books. Such books are written primarily for those who wish to study, not for those who desire to be intellectually entertained. But let me get through with my preliminaries, and then, under the guidance of science and photography, we shall try to visit the moon. One of the first questions that naturally arise concerning the objects that we see in the heavens relates to their distance from us. The average, or mean, distance of the moon from the earth is 238,840 miles. For the sake of a round number we usually call it 240,000 miles. But the orbit, or path, of the moon in her monthly journey around the earth, is so far from being a true circle that the distance is variable to the extent of 31,000 miles. Even the form of the moon's path in space is not constant. Owing to the varying effects of the attraction of the earth and the sun, her elliptical orbit becomes now a little more and now a little less eccentric, the consequence being that the moon's distance from the earth is continually changing. When she is at her greatest possible distance she is 253,000 miles away, but this distance at certain times, may be reduced to only 221,600 miles. As a result of these changes of distance the moon sometimes appears noticeably larger to our eyes than at other times.

"This leads us next to inquire, 'What is the actual size of the moon?' When we know the distance of any body from the eye it is not difficult to determine its size. The diameter of the moon is 2,163 miles. The face of the full moon contains 7,300,000 square miles. It is a little larger than the continent of South America. For a reason that we will speak of presently, the moon always keeps the same side toward us no matter in what part of its orbit it may be. Consequently we always see the same features of her surface and, except through inference, we do not know what exists on the other side of the lunar globe. Of the 7,300,000 square miles of surface which the moon presents to us, about 2,900,000 are occupied by those dark gray patches which you see so plainly spotting her face, and which were once supposed to be seas. The remaining 4,400,000 square miles consist of a very rough, broken country, ridged with gigantic mountains and containing hundreds of enormous craters, and mountain-ringed valleys, which are so vast that one hesitates to call them, what many of them seem evidently to be, extinct volcanoes. A single explosion of a volcano of the dimensions of some of these lunar monsters would shake the whole earth to its center!"

"Please stop a moment," my friend laughingly interrupted. "So many merciless facts, chasing one at the heels of another, are as bad as the books on your science that I have tried to read. Give my imagination time to overtake you."

"Very well," I said, "then relieve your attention a little while by regarding the face of the moon. Do you perceive the portrait of the Moon Maiden there?"

"I believe I do, although I never noticed it before. It is in profile, is it not?"

"Yes, and it occupies all the central portion of the western half of the disk. Take the opera glass and you will see it more clearly."

"Really, I find her quite charming," said my companion, after gazing for a minute through the glass. "But what a coquette! Look at the magnificent jewel she wears at her throat, and the *parure* of pearls that binds her hair!"

"Yes," I replied, "and no terrestrial coquette ever wore gems so unpurchasable as those with which the Moon Maiden has decked herself. That flaming jewel on her breast is a *volcano*, with a crater more than fifty miles across! Tycho, astronomers call it. Observe with the glass how broad rays shoot out from it in all directions. They are among the greatest mysteries of lunar scenery. And the string of brilliants in her hair consists of a *chain of mountains* greater than the Alps—the lunar Apennines. They extend more than 450 miles, and have peaks 20,000 feet high, which gleam like polished facets."

"Truly," said my companion, smiling, "these gigantesque facts of yours rather tend to dissipate the romantic impression that I had conceived of the Moon Maiden."

"No doubt," I replied. "It is only distance that lends her enchantment. But we must not disregard the facts. Her hair, you perceive, is formed by some of the vast gray plains of which I spoke a few minutes ago. She is like a face in the clouds—approach her, or change the point of view and she disappears or dissolves into something else.

"Now, to return to my preliminaries, upon which I must insist. Knowing the distance and the size of the moon, the next question relates to her motions. You are aware that she travels around the earth about once every month. There are two ways in which we measure the length of time that the moon takes for each revolution. First, regarding the face of the sky as a great dial, with the stars for marks upon it, we notice the time that elapses between two successive conjunctions of the moon with the same star. In the interval she has gone completely around the earth and come back to the starting point. This is called the moon's sidereal revolution, and it occupies, on the average, twenty-seven days, seven hours, forty-three minutes, twelve seconds. Every twenty-four hours the moon advances among the stars, from west to east, about 13° 11´.

"But there is another, more usual way of measuring the orbital period of the moon. This way is connected with her phases, or changes of shape, from the sickle of the New Moon to the round disk of the Full Moon, and back again to the reversed sickle of the waning moon. It is the time that elapses from one New Moon to the next, or from one Full Moon to the next which now concerns us, and it amounts, on the average, to twenty-nine days, twelve hours, forty-four minutes. This is called the moon's synodic revolution, and it is equivalent to the ordinary lunar month. It is variable to the amount of about thirteen hours. The reason why the synodic revolution is more than two days longer than the sidereal revolution is because the continual advance of the earth in its orbit around the sun causes the latter to move eastward among the stars, and before the moon's monthly phases, which depend upon her position with regard to the sun, can recommence, she must overtake the sun."

"What a hopeless task to try to remember all that!"

"At any rate, if you cannot remember these things my conscience will be clear, for I am simply doing my duty in telling you of them. If you forget, go to the books on astronomy and refresh your memory. But do not persuade yourself that the preliminaries are now finished. You are going to think that my story of the moon resembles Walter Scott's novels in the length of its introduction; but if, in the end, I can interest you half as much as he finally interests his readers I shall thank the stars for my good fortune.

"The next thing that I must try to explain," I continued, "is the cause of the moon's phases, or her continual changes of form. You know that the New Moon is shaped like a thin crescent, and always appears in the west immediately after sundown, with the convex side facing the setting sun. The moon at First Quarter is a half circle and is visible in the southern part of the sky just after sunset. The Full Moon, which we have at present, is a complete round disk, and is always seen directly opposite to the place of the sun, so that she rises when the sun sets. The moon at last quarter is again a half circle, and appears on the meridian in the south at sunrise. The waning moon is like the new moon, crescent-shaped, but the convexity of the bow faces the rising sun, and she is visible only in the morning sky just as dawn begins. To explain the reasons for these changes of shape, which the moon regularly undergoes every month, I must ask you to go indoors and examine a little diagram which I have made."

"Oh!" said my companion, "it is too bad to abandon this charming spectacle, illuminated by rays so fascinating, for the sake of looking at mathematical lines drawn on paper! But I suppose that this is one of the sacrifices demanded by your inexorable science, and must be made."

"Yes," I said, "but if science sometimes demands sacrifices, at least she always rewards them most generously."

When we had returned to the house I placed upon the drawing-room table this diagram.

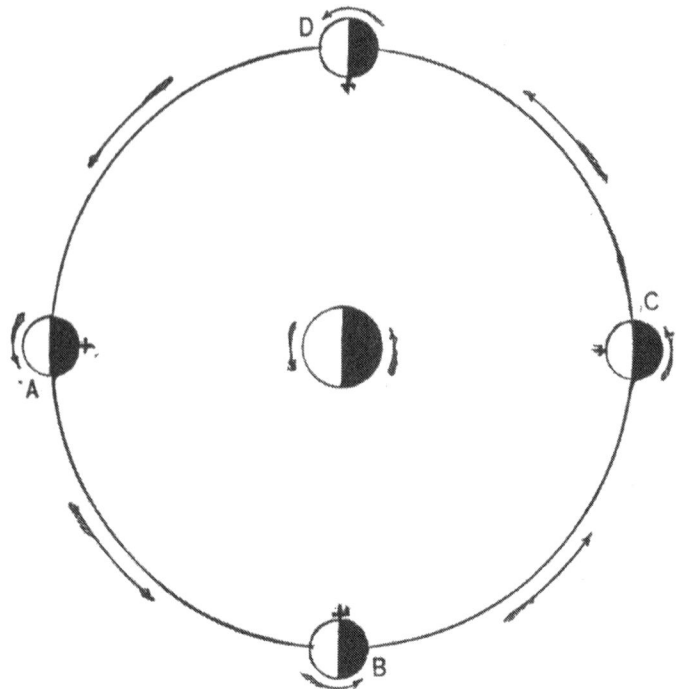

Phases and Rotation of the Moon.

As I spread it out, my companion, after a regretful glance through the open door at the silvery lawn, on which the moon, having cleared the obstructing branches of the bordering trees, was now pouring down the full splendor of her rays, put her elbows on the table to follow my explanation.

"The globe, half bright and half black, in the center," I said, "represents the earth. The large circle surrounding the earth we will call the moon's orbit, which she traverses once every month. The smaller globe, also half white and half black, shown in four successive positions in the orbit, is the moon. Suppose the sun to be away off here on the left. It illuminates the earth and the moon only on the side turned toward it. The opposite side of both is buried in night. Now, let us begin with the moon at the point A. She is then between the earth and the sun, the bright side being necessarily toward the sun and the dark side toward the earth. In that position we do not see the moon at all from the earth, unless she happens to come so exactly in a line with the sun as to cover the latter, in which event we have an eclipse of the sun. Now, suppose the moon to move in her orbit toward B. In a little more than seven days she will arrive at B. In the meantime, while moving away from the position of the sun, she begins to present a part of her illuminated hemisphere toward the earth. This part appears in the form of a sickle, or crescent, which grows gradually broader, until, at B, it has grown to a half circle. In other words, when the moon is in the position B we on the earth see one half of her illuminated surface. This phase is called First Quarter. The narrow crescent, which appears as soon as the moon begins to move from A toward B, is the New Moon. As the moon continues on from B toward C, more and more of her illuminated half is visible from the earth, and when she arrives at C, just opposite to the position of the sun, she becomes a Full Moon. We then see, as occurs to-night, the whole of that face of the moon which is presented sunward. The upper half of the diagram shows how the moon moves from the position of Full Moon back again to New Moon, or conjunction with the sun. During this latter part of her course the moon rises later and later every night, until, when she assumes the form of a waning crescent, she is visible only in the morning sky just before sunrise.[1]

"Now, there is another interesting thing shown by this diagram," I continued—but my companion, who had followed my explanations thus far with flattering attention, here suddenly ran to the door exclaiming:

"For mercy's sake, what is happening to the moon?"

1.

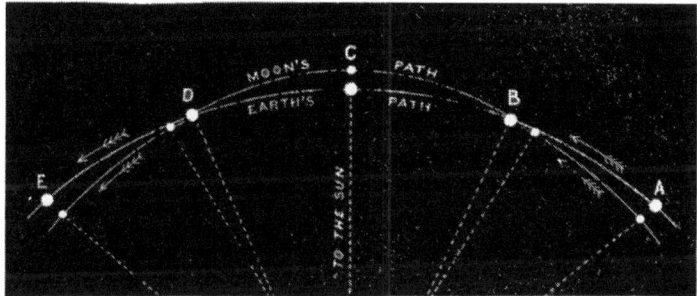

The Moon's Path with Respect to the Sun and the Earth.

It may be well to add to what is said in the text about the orbit of the moon, that, while the moon does perform a revolution around the earth once a month, yet her orbit is drawn out, by the common motion of both earth and moon around the sun, into a long curve, whose radius is continually changing, but which is always concave toward the sun. This is illustrated in the accompanying diagram. Suppose we start with the earth at A. The moon is then between the sun and the earth, or in the phase of New Moon. The earth's orbit at this point is more curved than the moon's, and the earth is moving relatively faster than the moon. At B (First Quarter) the earth is directly ahead of the moon. But now the moon's orbit becomes more curved than the earth's and it begins to overtake the earth. At C (Full Moon) the moon has come up even with the earth, but on the opposite side from the sun. From that point to D (Last Quarter) the moon gains upon the earth until she is directly ahead of it. Then, from D to E (New Moon, once more) the earth gains until the two bodies are in the same relative positions which they occupied at A. Throughout the entire lunation, however, notwithstanding the changes which the shape of the moon's orbit undergoes, the latter is constantly concave toward the sun. This shows that the sun's attraction is really the governing force, and that the attraction of the earth simply serves to vary the form of the moon's path, and cause it to move in a virtual ellipse with the earth for its focus.

I glanced over her shoulder, and saw a smudgy scallop in the moon's edge.

"Really," I said, "I am ashamed of myself. There is an eclipse of the moon to-night, and I had positively forgotten it! What you see is the shadow of the earth, which has the form of a long cone stretching away more than eight hundred thousand miles into space, and whenever our satellite at the time of Full Moon gets nearly in a direct line with the earth and the sun, it passes through that shadow and undergoes an eclipse. That is what is happening at the present moment."

"And the shadow has a round form because the earth is round, I suppose."

"Certainly; the shadow of a globe must have a circular outline. But the shadow of the earth, although it finally diminishes to a point, is, at the moon's distance, still about 5,700 miles in diameter, or more than two and a half times the diameter of the moon. In consequence of the motion of the earth in its orbit around the sun, its shadow constantly moves eastward, like a great pencil of darkness sweeping straight across the heavens, but invisible to us except when the moon, traveling eastward faster than the shadow, overtakes and passes through it. This does not by any means happen at every full moon, because, for a reason which I shall explain presently, the moon usually passes either above or below the shadow of the earth, and thus escapes an eclipse. When an eclipse does occur it lasts a long time because the shadow is moving in the same direction as the moon. The moon must pass entirely through it before the eclipse ends. On this occasion the moon will be in the shadow more than three hours, and during an hour and a half she will be totally immersed. We shall have plenty of time, then, to observe the phenomenon, and after you have satisfied your curiosity a little by watching the slow advance of the shadow movement across the moon, we can return to our diagram and finish its explanation before the eclipse becomes total."

Accordingly, after having watched the progress of the eclipse for half an hour, during which time the shadow began perceptibly to diminish the moonlight in the park, we returned to the lamplight and the diagram on the table.

"I was saying," I resumed, "that another interesting thing in addition to the cause of the moon's changing phases is represented here. You observe that a little cross stands on each of the four circles representing the moon, and that, in every case, the cross is in the center of that side of the moon which faces the earth. In fact the position of the cross upon the moon is fixed and invariable, and it always points toward the earth because the moon makes exactly one rotation on her axis in the course of one revolution around her orbit, or, as it is often called, one lunation. We know that this is so because we always see the same features of the lunar surface, no matter where the moon may be situated. This is true although, in consequence of the phases, we cannot see the whole face of the moon except when she is full. But whether it is the New Moon, or First Quarter, or Full Moon, or Last Quarter, or Old Moon, that we look at, the mountains and plains visible are identically the same. If the moon did not turn once on her axis in going once around the earth we would see all of her sides in succession, although only at Full Moon could we see an entire hemisphere illuminated by the sun. At Old and New Moon the side presented to the earth would be just the opposite to that presented at Full Moon. At Last Quarter the side facing the earth would be the opposite to that facing the earth at First Quarter."

"But, tell me," said my friend, "how did the moon ever come to so humiliating a pass that she must be forever turning on her heel to face the earth?"

"That," I replied, "is a result of the same forces which originally separated her from the earth and gradually pushed her off to her present distance. In a word it is due to 'tidal friction.' Before the moon had solidified, the attraction of the earth raised huge tides in her molten mass. These tides acted on the rotating moon like brakes on a wheel, and at length they slowed down her rotation until its period became identical with that of her revolution around the earth. For the mathematical calculations on which all this is based you must go to Professor Darwin's book on 'The Tides,' or some similar technical treatise; but I imagine you will never do that."

"Not just at present, I assure you. I do not know what unexpected ambition for the acquirement of scientific knowledge may arise after I have seen those wonders that you have promised to show me in the moon, but, for the moment, I am content to accept your statement of the simple fact."

"Good!" I replied. "And now, perhaps, you will have the patience to listen to an explanation of a very important relation which exists between the moon and the earth. We are led to it by what I have just said concerning tides. You know, of course that the tides in the oceans are due principally to the attraction of the moon. The sun also raises tides in the seas, but the moon, being so much nearer than the sun, is the chief agent in producing them. Sometimes the moon and the sun act together; at other times they pull in different directions. At Full Moon and at New Moon they pull together, because then they are either on opposite sides of the earth, or both on the same side. At such times we have the highest tides in all our seaports. That occurs about once every fortnight. But when the moon is at either First or Last Quarter, as you will perceive by looking at the diagram, her position, as seen from the earth, is at a right angle with a line drawn to the sun. Then the sun raises tides in one direction and the moon in another direction. The result is that at such periods the tides are lowest. An exact knowledge of these things is very important for mariners because there are harbors whose channels can be navigated by large ships only when the tides are high. Tables predicting the times and heights of the tides have been prepared for all the principal seaports of the world. In truth, the moon renders important services to the inhabitants of the earth, not merely in supplying them with a certain amount of light in the absence of the sun, but also in enabling them to navigate waters which are too shallow for ships except when deepened by the tide. The tides also, in many cases, serve to scour out channels and keep them open."

"Really, I am quite interested, and the more so because I find the moon, like a dutiful daughter, trying to be of some use to her mother. But have I not heard that the tides occur on both sides of the earth at once, and not simply on the side where the moon happens to be at the time? Please tell me how that can be so?"

"A complete reply to your question would carry us into the realm of mathematical physics, but perhaps I can throw a little light upon the matter with the aid of this second diagram.

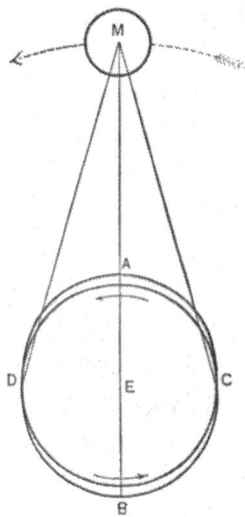

The Moon and the Tides.

"The eclipse is not yet total," I continued, glancing out of the door, "and we can finish our explanation before it becomes so. Have the kindness, then, to look at the diagram. Suppose E to be the center of the earth, and M the center of the moon. The protuberant portions of the earth C A D and D B C represent the waters of the ocean pulled away from the surface of the earth, if I may so describe it, by the moon's attraction. You are probably aware that the attraction of gravitation varies with the distance of the attracting body. The distance from the center of the earth to the center of the moon is about 239,000 miles. But the earth being nearly 8,000 miles in diameter, the surface of the ocean at A is about 4,000 miles nearer to the moon than is the center of the earth E. It follows that the force of the moon's attraction is greater at A than at E. If the water of the ocean were a fixed, solid part of the earth there would be no perceptible effect resulting from this difference in the amount of the moon's attraction. But since the water is free to move, to a certain extent, it yields to the attraction, and is drawn up a little toward the moon. At the same time it is, in effect, drawn away from C and D. The consequence is the production of a tide on the side facing the moon.

"Now, for the other tide, produced at the same time on that side of the earth which is turned away from the moon. The point B is about 4,000 miles farther from the moon than E; consequently the moon's attractive force is less at B than at E. From this it results that the body of the earth is more forcibly attracted by the moon than is the water at B. The earth therefore tends to move away from the water at that point, and another tidal protuberance is produced, with its highest part at B. I should add that while the water of the ocean is, to a certain degree, free to respond to these differences of attraction, the earth itself, being solid, can only move as a single body, and, mathematically, we may regard it as if its entire mass were concentrated at the center E. Please remember, however, that this explanation is only elementary, only intended as a graphic representation of the tides, and not as a mathematical demonstration of the way they are produced. Such a demonstration would only be suited to one of the technical books that you have not found as interesting as—some other branches of literature.

"There is just one other thing to which I must ask your attention, and then we shall return to the moon herself and the phenomena of the eclipse now in progress. You will notice in the diagram, that two arrows show the direction in which the earth is continually rotating on its axis, and that a dotted curve, terminating with an arrow point, indicates the course of the moon in her orbit surrounding the earth. The rotation of the earth is so much more rapid than the motion of the moon that the points A and B are carried out of the line drawn from the center of the moon to the center of the earth, in the direction of the arrows. From this it follows that the tides are never directly under the moon, or exactly opposite to her, but sweep in great waves round the globe. The tides produced by the attraction of the sun are only about two fifths as high as those caused by the moon. As I have already explained they are sometimes superposed upon the lunar tides—at New and at Full Moon—and sometimes they are situated at right angles to the lunar tides—at First and Last Quarters."

"But the eclipse!" interrupted my friend, whose attention had evidently begun to wander. "I think the totality of which you spoke must be at hand, for notice how dark the park has become, and the fireflies are so brilliant under the trees."

The total phase of the eclipse was, indeed, beginning, and we stepped out on the lawn before the door to watch it. The moon had now passed entirely within the earth's shadow, but although her light was almost completely obscured as far as its power to illuminate the landscape was concerned, still the face of the moon was dimly visible, as if concealed behind a thick veil. Certain parts of it had a coppery color, producing a very weird effect.

"Dear me!" exclaimed my companion, "I did not think it would look like that! I naïvely supposed that one could not see the eclipsed moon at all, but that she either disappeared or was turned into a kind of black circle in the heavens. And what a strange color she has! Positively it fills me with awe."

"It is very rare," I said, "for the moon to become invisible during an eclipse. That can only occur when the earth is enveloped in clouds."

"Indeed, and what have the clouds to do with it? If the solid body of the earth cannot cast a shadow dense enough to hide the moon, I should not expect things so evanescent as clouds to be more effective."

"It is all owing to the earth's atmosphere," I replied. "If our globe were not surrounded with a shell of air the moon would always be totally invisible when eclipsed. But the atmosphere acts like a lens of glass inclosing the earth; that is to say, it refracts, or bends the rays of sunlight around the edge of the earth on all sides, and throws a portion of them even into the middle of the shadow, at the moon's distance. It is these refracted rays which cause the singular illumination that you perceive on the moon. But when, as occurs only occasionally, all that part of the atmosphere which surrounds the earth along the edge visible from the moon is filled with clouds, the air can no longer transmit the refracted rays, and then, no light being sent into the shadow, a 'dark eclipse,' as astronomers call it, results. An eclipse of the sun is a very different thing. That is caused not by a shadow but by the opaque globe of the moon passing between the earth and the solar orb. When this occurs the sun is completely hidden behind the moon, and only its corona, which projects beyond the moon on all sides, is visible."

"Indeed! I supposed that all eclipses were very much the same thing."

"By no means. An eclipse of the sun is an event of extreme importance to astronomers, while an eclipse of the moon possesses comparatively little scientific interest."

"I do not see why that should be so."

"It is so, for the reason that when the sun is eclipsed, as I have just told you, the solar corona, which cannot be seen at any other time owing to the overpowering brilliance of the solar orb, becomes plainly visible, and by studying the form and other particulars of the corona astronomers are able to draw most important conclusions concerning the constitution of the sun, the mechanism of its radiation, and its effects upon the earth. During an eclipse of the moon, on the other hand, practically nothing new is revealed, and, accordingly, astronomers pay comparatively little attention to such phenomena. Lunar eclipses, however, possess a certain importance, because by predicting the times of their occurrence and then comparing the predictions with the events, something is learned about the motions of the moon. I should add that recently eclipses of the moon have been carefully watched by a few astronomers, notably by Prof. William H. Pickering, because of peculiar effects which seem to be produced at certain points on the moon by the chill which the shadow casts upon her surface. There are also interesting observations to be made concerning the reflection of heat from the moon during an eclipse. But, upon the whole, a lunar eclipse is mainly interesting as a curious spectacle, and as a test of the correctness of astronomical calculations of the motions of the heavenly bodies.

"I may add, however, that eclipses of the moon have been of some use to historians in fixing the dates of important occurrences thousands of years ago. This is possible because astronomers can by calculation ascertain the times of eclipses in the past as well as in the future. Perhaps the most interesting of all instances of this kind is that which relates to the date of the beginning of the Christian era. This has been fixed by means of an eclipse of the moon mentioned by the ancients as having happened the night before the death of Herod, king of the Jews."

"It seems to me," said my friend, "that the faint light on the moon's face is continually changing. It does not appear constantly to have the same tint. While we have been standing here, I have noticed that some parts have grown darker and others lighter, and that the red color on the right has become a little more intense."

"Yes, and that, too, is no doubt caused by the earth's atmosphere. While the eclipse lasts, the earth is rapidly rotating, and consequently new parts of the atmosphere are continually brought to the edge where their refractive effects come into play. If the atmosphere at the edge of the earth is a little more or a little less dense its refraction varies proportionally. Then, changes in the relative clearness or cloudiness of the air are taking place all the time, and these are reflected in the illumination on the moon."

"It seems to me, then, that the earth would present a very remarkable spectacle if we were now on the moon looking at it."

"Surely it would. Seen from the moon the earth appears several times larger than the sun. For the people of the moon, if we imagine them to exist, an eclipse of the sun is now in progress. For them the earth now occupies the same relative position which the moon occupies for us just before it appears in the west as New Moon. They cannot see it except in silhouette as it passes over the sun. More than an hour ago, if they were watching (and if they exist, and are intelligent beings we may be sure that they were on the alert), they suddenly perceived a black round-edged notch in the side of the sun. Instead of being more or less cloudlike and indefinite in outline, like the shadow of the earth on the moon, this notch appeared to them perfectly black and smooth. At a glance, they saw that the body producing it was much larger than the sun. As the sun's disk was gradually hidden behind the earth the shadow of the latter fell over them, until the sun was wholly concealed. Then—and this is true at the present moment—they perceived that the huge disk of the earth was ringed with light, probably of a reddish tinge. This light, as I have already indicated, is that which the atmosphere refracts around the edge of the earth."

"It must be truly a magnificent sight," said my companion.

"Yes, and it is doubtless rendered far more magnificent by the other phenomena which our people at the moon have before their eyes. In consequence of the virtual absence of air there, an observer on the moon would see all the stars, even in full daylight, blazing in a jet black sky. The brilliance of the stars and of the Milky Way would hardly be increased by the hiding of the sun, but probably the long silvery streamers of the solar corona would glow perceptibly brighter when seen projecting out on each side of the enormous disk of the earth."

"But is it true that the moon has no air?"

"Very, very little, and what little she has is probably different in composition from our atmosphere. Some observations seem to indicate that there is a very rare atmosphere on the moon, but to us it would seem a perfect vacuum. We could not breathe there at all."

"How then do those intelligent inhabitants, whom you have pictured for me watching the earth at this moment, manage to survive?"

"Ah, I did not say that there actually are inhabitants in the moon. I only imagined them to exist for the sake of showing how this eclipse would appear seen from the moon. Still, we cannot be absolutely sure that there are no inhabitants on the moon. Even without air like ours it is conceivable that beings of some kind, and intelligent beings, too, *might* exist there. However, astronomers have never yet been able to discover evidence of their presence. Lately, indications have been found of the probable existence of vegetation on the moon, but I shall speak of that later, when with the aid of the series of lunar pictures made at the Yerkes observatory we try to make a 'photographic journey' in the moon."

"But tell me, has the moon always been so airless?"

"That is another unsettled question. Some astronomers have thought that formerly, ages ago, the moon possessed a much more dense atmosphere than she has at present. Having separated from the earth, in the way I have described, it is natural to suppose that at first she may have had an atmosphere very like ours. The explanation of its disappearance which was once generally accepted was that it had been absorbed into the lunar rocks, as the globe of the moon cooled off. But recent progress in our knowledge of the nature of the gases composing the atmosphere has led to a different explanation. This assumes that nearly all of the moon's atmosphere has *flown away from her* because the lunar globe does not possess sufficient gravitating force or attraction to retain it. If the mass of the earth were no greater than that of the moon, our atmosphere also would probably have escaped by flying off into space."

"But how, and why, do these gases fly away?"

"They do it by virtue of what physicists call their molecular velocity. A gas, of whatever kind, is a mass of molecules which are in continual vibration, moving in all directions among one another with very great velocities. These velocities have been measured, and it has been found that the molecules of nitrogen, one of the components of the air, move at the rate of two miles in a second. The velocity of the molecules of oxygen is a little less; that of the molecules of hydrogen is very great, nearly seven and a half miles in a second! Now, it is also known that the attraction of the earth is sufficient to retain permanently upon its surface all moving particles or molecules which have a velocity less than seven miles in a second, while the attraction of the moon only suffices to retain those whose velocities fall under a mile and a half in a second. So you perceive that all of the gases I have named would soon escape from the moon, even if they were present upon it at the beginning of its history.

"I must also remind you that there is no water upon the moon, at least not in the form of rivers, oceans, lakes, ponds, or even of clouds. But Professor Pickering has recently noted certain appearances which may be due to the formation of a kind of hoar frost. If there were once oceans upon the moon, as the great plains, called *maria*, or seas, in the lunar charts, seem to indicate, they, too, have escaped by evaporation. The velocity of the molecules of water vapor is two and a half miles per second, a mile greater than the 'critical velocity' which the attraction of the moon would be able to control."

"But," interrupted my companion, "I am puzzled to understand how you know so much about the power of the moon to hold things."

"It is really quite simple," I replied. "The attraction of gravitation, which is a property belonging to all known bodies, is measured by the mass, or amount of matter, in a body. It also varies with the distance between the attracting and attracted bodies. We know, by means which I shall not attempt to describe here, the mass both of the earth and of the moon. We also know the size of both of these bodies. They attract objects as if their entire masses were concentrated at their centers. A body of a certain kind and size at the surface of the earth weighs just one pound. If the earth were reduced to half its actual diameter, while retaining the same mass or amount of matter, more closely packed together, the body which now weighs one pound would then weigh four pounds, because it would be twice as near to the center of the earth as before, and the attraction of gravitation varies according to the square of the distance from the center. As the distance diminishes the force increases. The square of two is four, therefore the body would be attracted with four times the force which it experiences at present. Now, the moon is not only much smaller than the earth, but its average density, or the closeness with which the molecules of its rocks are packed together, is less. It results from these facts that the ratio of the entire mass of the moon is to that of the earth as one to eighty-one. Hence the inherent power of the moon to attract bodies is less than one-eightieth as great as the earth's. If the diameter of the moon were the same as that of the earth, a body weighing one pound on the earth would weigh only one eighty-oneth part of a pound on the moon. But the diameter of the moon is less than one quarter as great as that of the earth. It follows that bodies on the moon are almost four times (more accurately about 3.66 times) nearer to the center of attraction. This fact must be taken into account in calculating the force of gravity on the moon's surface. As far as the mass of the moon is concerned, bodies on her surface experience less than one-eightieth of the attractive force which the earth exercises upon bodies on its surface, but this is so far counterbalanced by their greater nearness to the center, that the actual attraction upon them is about one sixth of that which they would experience on the earth."

"Thank you," said my companion dryly, "your explanation appears to me to be very scientific."

"Not by any means as scientific as it might be, or as it ought to be," I replied, laughing. "But, really, if you wish to understand these things you should not be too much afraid of the bugbear 'science.' Science makes the world go nowadays, and everybody ought to know a little about it, just as everybody with any pretensions to education a hundred years ago had to learn more or less Greek and Latin. But let me continue a little farther. Since the force of attraction on the moon is only one sixth as great as it is on the earth, the weight of all bodies is in the same proportion. Pardon me if I guess at your weight; it is, perhaps, 120 pounds. Very well, translated to the moon you would weigh only 20 pounds."

"Dear me, then skipping the rope may be the favorite pastime of middle-aged ladies on the moon."

"And throwing somersaults that of gray-haired lunar gentlemen. Let me tell you of one very interesting consequence of the small force of the moon's gravity, which affects not merely the weight of bodies but the flight of projectiles, and, indeed, all motions of every kind. You will see, when we come to the photographs, that some of the lunar volcanoes are of a magnitude almost incredible. This is doubtless due to the fact that the ejections from volcanic craters there were able, with no greater expenditure of explosive force, to attain an elevation six times that which they would attain if thrown from a volcano on the earth. During the eruption of Vesuvius in April, 1906, the column of smoke, steam, and cinders from its crater reached, according to the measures of Professor Matteucci, a maximum height of about eight miles. On the moon the same force would have blown these things almost fifty miles high! It is not difficult, in view of such facts, to see how the giant volcanic craters and mountain rings of the moon were formed."

In the meantime the eclipse continued, and, having tired of watching it, we returned to the drawing-room.

"When shall we see these famous photographs and begin our imaginary journey in the moon?" my companion asked.

"To-morrow," I replied. "But I shall have to demand one more brief exercise of your patience this evening, while I finish with this subject of eclipses."

"Then we are not through yet?"

"Not quite, for I have not yet told you why the moon is not eclipsed every time she approaches the earth's shadow, and why she does not eclipse the sun once every month at the time of New Moon."

"Well, tell me then, and I promise to be as interested as possible; only please don't talk any more mathematics than is absolutely necessary."

"Very well, I'll spare your attention as much as possible. To begin with the eclipses of the moon: The reason why they are not of regular monthly occurrence is simply because the orbit of the moon is a little inclined, about 5¼°, to the orbit of the earth. Even then there would be an eclipse once every month if the orbit of the moon were fixed in space, and if the point where that orbit crosses the plane of the earth's orbit were always directly opposite to the sun. But instead of being fixed in position the orbit of the moon has a curious motion of revolution of its own. This causes the two opposite points, where it crosses the plane of the earth's orbit, and which are called the moon's 'nodes,' to move continually onward in a direction opposite to that in which the moon revolves, but much more slowly. A period of about nineteen years is required for the moon's nodes to complete a revolution. The consequence is that the nodes are not always in line with the earth and the sun, and except when they *are* nearly in line no eclipse can occur. To enter into a complete explanation of this would require more 'mathematics' than you would like, but what I have said may at least serve to give you an idea of the reason why eclipses are comparatively of rare occurrence."

"I think I understand the reason sufficiently. But what a complicated affair you astronomers make of what, it seems to me, should really be a very simple thing."

"It is like a sewing machine," I replied, "which seems very simple when you see it running smoothly, and do not trouble yourself about all the various parts of its mechanism. But if you undertake to explain to yourself, or to make clear to another person, exactly how the machine works, you find that your attention is rather severely taxed, and that the apparent simplicity is based upon no little complexity of construction and interaction of parts. You will have understood from what I have said, that the reason why the moon does not eclipse the sun once every month is based upon the same fact, namely, the inclination of the moon's orbit to the plane of the orbit of the earth; and that when she does eclipse the sun her nodes must be somewhere near a line drawn from the earth to the sun. There is one broad difference between an eclipse of the moon and an eclipse of the sun which I have not yet mentioned. This arises from the fact that the moon being so much smaller than the earth, her shadow, when she hides the sun, does not cover the entire earth, as the earth's shadow covers the whole moon, but comes almost to a point before reaching the earth. The average length of the moon's shadow is only 232,150 miles,

6,690 miles less than the average distance between the moon and the earth. But since, in consequence of the eccentricity of her orbit, the moon's distance is continually varying, the length of her shadow also varies to the extent of about 4,000 miles each way. Thus it may be as short as 228,300 miles, or as long as 236,050 miles. When the greatest length of the moon's shadow coincides with her least distance from the earth (221,600 miles), her shadow extends more than 18,000 miles beyond the earth. Under such circumstances its diameter at the surface of the earth is about 167 miles. That is the greatest diameter that the shadow of the moon can have at its intersection with the earth. Ordinarily, when it reaches the earth at all, its diameter is less than 100 miles, and often very much less. If the earth and the moon were motionless during an eclipse, her shadow would form a round, dark spot on the earth, and all observers within the circumference of that spot would behold the sun totally eclipsed. But, in consequence both of the motion of the moon in her orbit, and the rotation of the earth on its axis, the shadow spot moves swiftly in an easterly direction over the earth's surface, forming what is called the path of the eclipse. The astronomer calculates beforehand across what parts of the earth the path will lie, and selects his points of observation accordingly.

"When the length of the shadow is too small to reach the earth, the moon appears projected against the sun as a round black disk, hiding the center of the solar orb, but leaving a brilliant ring all around. Such phenomena are called annular eclipses. There are about three annular eclipses for every two total ones. When the moon, as often occurs, does not traverse the center of the sun's disk, as seen from any part of the earth, a partial eclipse is the result. This means that only a portion of the sun is hidden by the moon. Even a total eclipse appears as a partial one to observers who are not placed within the limits of the shadow path."

"But it seems to me," said my friend, "you have hedged round your eclipses with so many difficulties, what with the tip of the moon's orbit, and what with the shortness of her shadow, that they must be very few in number. Yet I often hear of an eclipse, although I have never seen one before to-night."

"They are not so rare as you might suppose," I replied. "It is not necessary, in order that an eclipse, either partial, or total, or annular, may occur, that the moon's nodes be in a *direct* line with the centers of the sun and the earth. The moon may be a few degrees out of line, and yet either pass into the earth's shadow or be seen crossing the sun from one point or another on the earth. There are about 70 eclipses in every eighteen years, 41 of the sun and 29 of the moon, but the number varies a little. Generally there can be no more than two eclipses of the moon in any one year, but it is possible for three to occur. The greatest number of solar eclipses in a year is five, but this is very rare, the usual number being two. In fact, there must be at least two solar eclipses in a year, but there are many years which have no eclipses of the moon at all. And now, I think I have said all that is necessary about eclipses, and we arrive very opportunely at the end of the discourse, for behold the moon is passing out of the shadow, and her light begins once more to glow in the park."

This was indeed the case. Going to the door, we saw the earth's shadow slowly withdrawing from the face of the moon, while the landscape was brightening under her returning rays. For a few minutes we watched, in silence, the brilliant spectacle. Then my companion turned to me.

"Would you know my whole thought?" she asked. "I fear that I cannot recall many of the scientific facts you have just been telling me, but for them I can go back, at need, to the books. Yet one thing I feel that I have certainly gained. It is a sense of friendly, companionable interest in the moon. Henceforth she will be more to me than she ever was before. I shall always be conscious, when looking at her face, that she is the offspring of the earth, and that there exists between these two bodies an intimacy that I had never imagined possible. For me your tides and your eclipses seem an inarticulate language, a caressing exchange of communications between these two celestial beings of one blood. To my mind they are, in a certain sense, personalities, and, as a creature of the earth, I feel now my relationship to the moon."

"Very good," I replied. "All science and all forms of knowledge are rooted in the imagination. To-morrow we shall begin with the photographs, and many most interesting things that I have not yet mentioned will then naturally present themselves before us."

"Good night then," said my companion, "and to-morrow I shall count upon the delights of a photographic journey in the moon."

I NEW MOON TO FIRST QUARTER

I NEW MOON TO FIRST QUARTER

AT breakfast the next morning I asked my friend if she still had sufficient curiosity concerning the moon to induce her to undertake the contemplated journey amid lunar scenes.

"Yes, surely," she replied. "My dreams last night were filled with wonderful spectacles; great cones of shadow flitted continually through the heavens, eclipsing, in turn, moon, sun, and stars; and I stared, as it seemed, for hours at strange faces veiled behind a maze of mathematical diagrams covering the moon. I am not sure that your discourses have made me scientifically much wiser, but I feel that my imagination is sufficiently aroused to enable me to enjoy the photographic excursion that you have proposed, and I am quite ready to start at once."

"Excellent!" I said, producing my portfolio. "Here then are the photographs which I trust will enable us, in imagination, to spend an interesting month upon the moon. These photographs were made at the Yerkes observatory and they represent the moon, as you will perceive, in all of her principal phases, beginning with the narrow crescent of the New Moon, and ending with the similar, but reversed, sickle of the Old Moon."

"Let us take them out into the park under the trees," my friend suggested.

The shafts of morning sunshine, falling through the branches and illuminating the broad lawns and brilliant flower-beds, offered the greatest possible contrast with the strange scenes of the preceding night. We chose the shadow of a huge elm, and had a table placed there for our accommodation. On this I spread the photographs, and my companion began to examine them with many expressions of interest.

"It is not often," I said, "that science finds so flattering an audience."

"And I hope, surely, never so small a one," she responded, laughing. "But you must admit that science very seldom presents herself in so attractive a form as that of these pictures."

"They are indeed of the highest excellence," I replied. "It is the very moon herself that you see there."

"But are you certain that they have not been embellished? Has not the hand of an artist retouched and improved them—particularly these large ones that seem to contain a thousand curious things which I can hardly believe really exist on the moon?"

"No," I said, "there is nothing fictitious or imaginary in what you see. The only art displayed here is that of the astronomer-photographer, whose greatest ambition is to make his pictures absolutely true to nature. A defect in one of his plates, producing the appearance of a speck of light or shadow which does not actually exist, causes him as much distress of mind as you would experience upon hearing a false note from your piano. Indeed, the astronomer is so desirous of having nothing but the truth represented in his pictures that he often prefers, for his own study, the original negatives alone, because every time that they are reversed to make a 'positive' copy something is sure to be lost, and some slight defect is certain to be introduced. Let us begin, if you please, with the series of smaller pictures showing the various phases, and the gradual advance of daylight across the moon's surface. Take first the photograph which I have labeled No. 1. It shows the New Moon when it is between three and four days old. You must often have seen it in that form in the western sky soon after sunset. Photographs of the New Moon have been made when the crescent is still narrower than that here shown, but there is no such photograph in this series, and it would possess little interest for you because almost no details of mountains, craters, and plains would be visible. It is hardly possible to make a good photograph of the moon when it is only one or two days from the sun in its monthly journey, on account both of the glare of the solar light in our atmosphere and of the nearness of the moon to the horizon, where the air lacks transparency and steadiness. In the photograph before us you will observe a great number of strange forms and shadings. I shall tell you what these are presently, but first let me call your attention to the fact that the picture does not exhibit a phenomenon which you would behold if you were actually looking at the moon in the phase here represented. You see here the New Moon very clearly, but not the Old Moon in her arms."

"Indeed! It is a pity that the photograph does not show so interesting a sight."

"Yes, it is a pity. The cause lies in the defect of light from what I have called the 'Old Moon.' The part that we see in the photograph is illuminated with sunshine, while the remainder of the moon reflects only the earthshine, which is too faint to be photographed (at least with the amount of exposure required to make a good picture of the brightly lighted crescent); although, as I have said, you would see it clearly if you were looking at the New Moon herself."

No. 1. February 19, 1904; Moon's Age 3.85 Days.

"But," interrupted my companion, "do you mean to tell me that the earth illuminates the moon?"

"Surely it does. Why not?" I replied, smiling. "You must remember that the earth is simply a huge moon to our imagined inhabitants of the lunar world. Our globe sends to the moon about fourteen times as much reflected sunlight as the moon sends to the earth. The consequence is than an earthlit night on the moon is far more brilliant than a moonlit night on the earth."

"Then why do we not always see the moon shining with light from the earth?"

"It is a question of contrast. You cannot see a faint light in the immediate presence of an overpoweringly brighter light. The part of the moon that the sun illuminates is in the full glare of day, and this is so much more brilliant than the reflected earthlight that that portion of the moon which enjoys only the latter is not visible to us, except for a few days after New Moon, when the amount of light from the crescent is not yet great enough to dazzle our eyes and hide the rest from sight. I should advise you when the next New Moon occurs—you can find the date in any almanac—to look at it in the western sky. You will see in addition to the bright crescent the full round orb of the moon, shining faintly, with a dull, rather copperish, tint, and you will find it interesting, then, to remember that that light is reflected from our earth.

"And now," I continued, "let us examine our photograph more closely. There is one remark that I had expected which you have not made; it concerns the position of the crescent. You observe that it is bowed toward the left. If you saw it with the naked eye in the sky it would be bowed toward the right, or toward the place of sunset. The reason is that the photograph presents the moon as seen with a telescope, which reverses objects, turning them top for bottom. In this picture, and in all the others that we shall examine, the southern part of the moon is at the top and the northern part at the bottom, the western part at the left and the eastern part at the right. The first thing that you probably notice in the photograph is a conspicuous oval plain, somewhat below the center of the crescent."

"Yes, and I see clearly why you call it a plain, for it is perfectly flat and smooth."

"Not quite so flat and smooth as you suppose. This object is one of the most celebrated on the moon. It is the so-called *Mare Crisium*, or Sea of Crises, as we may translate the name given to it by the astronomers of a couple of centuries ago, many of whom knew more Latin than science. Owing to its apparent smoothness of surface, as well as to its form and general aspect, they took it for a great lake or sea."

"To tell you the truth," said my friend, "if I were an astronomer and had discovered this curious place on the moon, I should certainly believe just what your Latin-loving predecessors believed, but I doubt if I should have been capable of inventing so singular a name for it."

"In the singularity of the names they chose for objects on the moon," I replied, "their invention is unrivaled. We shall see some remarkable examples. Of course they are not at all to be blamed for thinking that this oval spot, and other similar ones of much greater magnitude, were seas and oceans. They simply judged by appearance and by analogy. Finding mountains on the moon, they saw no improbability in supposing that there were bodies of water also. They had not the means of knowing, as we know to-day, that there is no water on the moon. Yet, perhaps, they were not so far wrong after all. The *Mare Crisium* certainly has the look of an empty sea bed, and I should not be willing to assert that ages ago it was not filled with water."

"Like the Great Salt Lake, dried up," suggested my companion.

"Not exactly, for the Great Salt Lake dried up would probably present a surface as white as snow, whereas the *Mare Crisium* is very dark. It must be admitted, however, that gradually the white deposit would grow darker, and there may be much significance in the fact, which some observers have noticed, that, at times, parts of the dark plains on the moon seem to glitter with minute points of light. Your imagination is at liberty to see deposits of salt there."

"In that case," said my companion, laughing, "I should prefer to regard the *Mare Crisium* as resembling that wonderful valley discovered by Sindbad the Sailor, whose floor was sprinkled with diamonds."

"Well," I replied, "science certainly cannot deny the possibility of diamonds on the moon, for she is *par excellence* the world of volcanoes, and one of the most striking discoveries of recent years is that of the intimate association existing between ancient volcanic vents and deposits of diamonds. The diamonds of South Africa are found in lava rocks that cooled off ages ago."

"Then I hope that no future Columbus will find a way to the moon, for we should have too many diamonds, and they would lose all their charm."

"That is true, but suppose that not only diamonds but even more beautiful gems should be discovered in the lunar world? You surely would not object to a transethereal traffic bringing them to our doors. However, there is not the slightest prospect that we shall ever be able to go from the earth to the moon. Let us resume our examination of the photograph, and concentrate our attention on the known facts."

I then proceeded to tell my friend, whose interest I was delighted to find had not yet begun to flag even in the face of comparatively matter-of-fact statements, that the *Mare Crisium* is a profound depression, about 350 miles in length by 280 in breadth. Exactly how far it lies below the general level of the lunar surface we do not know; but, at any rate, if it was ever filled with water it formed a deep, navigable sea. Its encircling mountains, which appear generally bright in the photograph, especially along the eastern border, where the sunlight strikes directly against their slopes, are in many places steep and abrupt. At one place, on the southwestern side, there is a mountainous promontory 11,000 feet in height. There are a number of small craters on the floor of the *Mare Crisium*, but the scale of this photograph is not large enough to show them clearly.

"You will notice," I continued, "that there is a kind of bay on the eastern side, which runs back into the mountains, and is bordered with high, steep cliffs. Near this point, on that part of the moon over which the sun has not yet risen, there is a very remarkable mountain which we shall see in a later photograph. But let us finish with this one. Look at the comparatively small oval adjoining the *Mare Crisium* below (toward the north). It is one of the great crater rings of the moon, and is named Cleomedes. It is much larger than it looks, being nearly 80 miles in its greatest diameter, and there is a peak on its surrounding wall 10,000 feet in height. Still farther toward the north you will observe two or three other smaller craters or rings, which are very interesting when studied with the telescope.

"Now, please turn your attention to the photograph bearing the number 2. You see again the *Mare Crisium*, and nearly in the center of the crescent, and just on the border line between day and night, a perfect oval ring with a central peak. It is called Langrenus. It is even larger than Cleomedes, being about 90 miles across. It has the form of an oval, as we see it, but that is an effect of perspective, since it is so far round the side of the lunar globe. In reality it is a nearly circular circumvallation, or rather an almost perfect hexagon, composed of gigantic mountains including a valley, in the center of which rises a cluster of peaks 3,000 feet in height."

"This second photograph," interrupted my friend, "was taken later than the first, I suppose, since it shows more of the moon's surface."

No. 2. September 24, 1903; Moon's Age 3.87 Days.

"I should have told you that," I replied. "Yes, it does represent the moon at a time when more of its surface, visible to us, is illuminated by the sun. In fact, we may regard it as a picture of the moon made about a day later than the other. But I must now tell you that these photographs were not all taken in regular succession, a day apart, or even two days apart. That was impracticable for reasons that I need not explain. Some of them were made at one season of the year and some at another. Yet taken together they form a sufficiently continuous series to enable us, with their aid, to follow the changing aspects of the moon during more than three weeks, or all that part of a lunation in which the moon is a conspicuous object in the sky."[2]

2. In addition to what is said in the text concerning the photographs the reader should be informed that, in consequence of her "librations," the moon does not, all the time, present *exactly* the same surface toward the earth. If she did we should never see more than one half of her surface. In fact, however, at one time or another, we see, in all (but never at the same time),

about fifty-nine per cent of her surface, leaving forty-one per cent which is forever invisible because never turned in our direction. The librations, or "balancings," of the moon, which bring now one and now another portion of the usually invisible hemisphere into view, are of three kinds: First, the libration in latitude, arising from the combined effects of the inclination of the moon's orbit to the plane of the earth's orbit, and the inclination of her axis of rotation to the plane of her own orbit. When added together these two inclinations make the axis of the moon lean one way or the other with respect to the earth about 6½°. But, since the inclination of the moon's orbit to that of the earth is continually varying to a small extent, the amount of this libration is also variable. Its effect is to cause now the North and now the South Pole of the moon to incline slightly toward the observer on the earth, so that he can see alternately a little way round the northern and the southern edges of the moon's disk.

Second, the libration in longitude, which arises from the eccentricity of the moon's orbit, causing her to move a little faster when she is nearer the earth, or in perigee, and a little slower when she is farther from the earth, or in apogee. In consequence of this, she gets alternately about 6° ahead of, or behind, the position which she would have if her orbit were a perfect circle and her motion perfectly uniform. But, inasmuch as her rotation on her axis is never either faster or slower, she shows a little of her usually invisible hemisphere on the western side when she is between perigee and apogee, and a little on the eastern side when she is between apogee and perigee. The accompanying diagram is designed to aid the reader in understanding these effects.

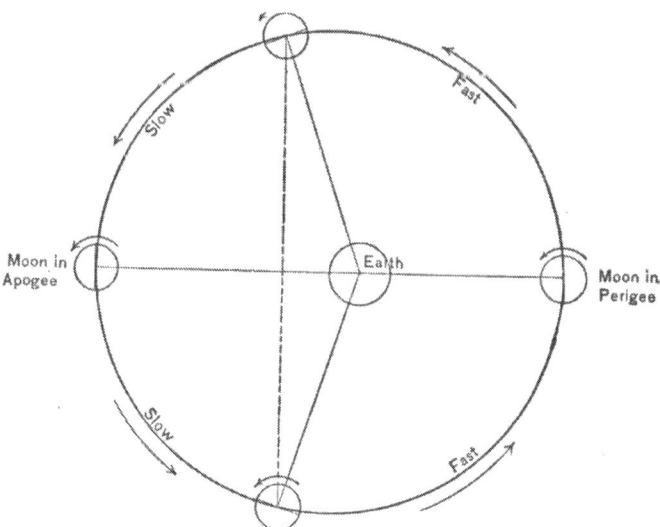

Effect of Moon, Varying Velocity in Orbit Producing Libration in Longitude.

Third, the diurnal libration, which arises from the fact that the diameter of the earth bears a considerable proportion to the distance of the moon. If the observer were at the center of the earth there would be no effect of this kind, but being situated about 4,000 miles from the center, there is a parallactic effect in consequence of which we see a little around the western side of the moon when she is rising and a little around the eastern side when she is setting. The maximum diurnal libration is a little more than one degree. The maximum libration in latitude is 6° 44′, and that in longitude 7° 45′. An illustration of the results of libration will be found by comparing photographs Nos. 1 and 2. They were both taken at nearly the same "age of the moon," about three days, twenty hours, but under different librations, so that in No. 2 more of the western edge of the moon is visible, and the crescent appears broader. Even more remarkable examples of the results of libration are seen in Nos. 6 and 7, and 8 and 9. In No. 6, the moon is actually "older" by about half a day than in No. 7, yet, owing to libration, the "terminator," or line between day and night on the moon, is considerably farther toward the east in the latter than in the former. A similar effect is seen in comparing Nos. 8 and 9. The exact dates and ages of the moon corresponding to these photographs are given in the Appendix.

"If you will follow the curve of the terminator toward the south (upward in the photograph), you will perceive that there is a long line of ovals, more or less resembling Langrenus. The first of these, darker in appearance than Langrenus, is named Vendelinus."

"What extraordinary names!" exclaimed my companion, "and how unpicturesque!"

"Yes, it is true that the invention of the old astronomers who supplied these names seems to have failed a little at times. They did exceedingly well in naming the 'seas' and similar objects, but for the mountains, craters, and ring plains they could think of no better plan than that of attaching to them their own names, and the names of other *savants*, or supposed *savants* of their time, or of preceding centuries. And in Latinizing these names they gave them a kind of uniformity, which is hardly pleasing to our taste to-day. But let me continue. Vendelinus is an extremely beautiful sight when the sunlight strikes its broken walls in such a manner as to bring into prominence, by contrast with the deep shadows, the rugged peaks, precipices, and ridges of which its very irregular ring is composed. You should see it with a powerful telescope, especially under the rays of the setting sun. Then the bottom of the valley within has been described by Mr. Eiger, an English student of lunar phenomena, as appearing punctured like a sieve with holes."

"And what are they?"

"Volcanic craters, probably, long since extinct."

"So many volcanoes in one place?"

"Oh, yes. You have been at Naples and have seen Vesuvius. But probably you have not visited the Phlægrean Fields which lie northwest of Naples. If you had had a passion for geology when you were in Italy you would have explored that region, and there you would have found something not altogether unlike the valley of Vendelinus in the moon. There is a great number of extinct volcanic craters near Naples, and they show how similar in many ways the moon is, or has been, to the earth."

"But, dear me," my friend exclaimed, "are we going to see nothing but burned-out craters and wild, ragged mountains on the moon? I am sure that I should never have thought of visiting Naples for the sake of looking at its Phlægrean Fields."

"Still," I replied, "you must certainly know that Pompeii and Herculaneum and the memories of their tragic fate are the most vivid attraction of Naples to-day, although the Pompeiians have all been dead for almost 2,000 years. So in looking at these spectacles in the moon we cannot but be interested by the reflection that they are reminders and relics of a wonderful history, whatever its precise character may have been. The moon seems to me to stand for the most affecting of all tragedies—the passing of a world. When I survey its extraordinary landscapes, it is like looking upon a long-abandoned stage, whose actors are in their graves, whose scenery is moldering under a gaping roof, whose machinery is broken, whose very traditions are forgotten, but which yet retains a semblance of its former brilliance. I do not have to imagine inhabitants in the moon at the present day in order to find it interesting. The possibility that it may once have had inhabitants is enough, remembering its nearness to the earth and the manner of its origin, to make it the most fascinating thing that the heavens contain."

"Indeed, I had never thought of the moon quite in that way," was the reply. "If you can read a history for me in these craters and ring plains I believe I shall find them more interesting than I expected."

"I cannot promise you a history as full of romantic details as that of Herodotus," I said, "but it may contain nearly as many actual facts. However, we shall see about that as we go along. Let us now return to the inspection of the photograph. Be kind enough to look a little above Vendelinus. You observe there another still larger ring plain, or walled valley, with a conspicuous mountain in the center. This is Petavius. It belongs to the chain of similar formations which includes Langrenus and Vendelinus, but it is more wonderful than either of them. It is nearly a hundred miles long from north to south. For some reason, as with Vendelinus, its ruggedness and complexity of structure are more conspicuous in the lunar afternoon than in the lunar morning. It is a question of the direction in which the light falls across it. A curious thing about Petavius is the convexity of its vast floor. The center is about 800 feet higher than the edges along the feet of the surrounding mountains."

"How do you know that?"

"The shadows tell the story. The height of objects on the moon is measured by observing the length of their shadows under a known inclination of the sun's rays. When I stand this book upright on the table, allowing the sunlight to strike it on one side, it casts a shadow on the table. If I did not know the height of the book, and could not measure it directly, I could find it out by measuring the length of its shadow, other simple trigonometrical data, easily ascertained, being known. There is an enormous cleft not clearly visible in the photograph, extending from the central mountains of Petavius to the southwestern wall of the valley. Still farther south, above Petavius, you will notice another conspicuous oval plain and several smaller ones near it. The largest of these is named Furnerius. They all lay in the morning sunshine, not far from the terminator, when this photograph was taken."

"Tell me, please, about the 'terminator' of which you have spoken several times. As I understand you it is the line between day and night on the moon."

"Yes, and a very wonderful line it is, too. There is nothing just like it on the earth. Owing to the effects of our atmosphere in dispersing the light, day and night do not stand face to face with one another on the earth in the same way that they do on the moon. Here we have twilight in the evening and dawn in the morning, and night neither comes nor goes for us with the startling suddenness that characterizes it on the moon. For an hour or two after sunset and before sunrise, we receive rays of reflected and refracted light from the atmosphere above us, which spread a soft, pleasing illumination over the landscape, and render all objects more or less distinctly visible. But if you were on the moon in certain situations, the passage from day to night or from night to-day would be as rapid as the falling or rising of a curtain. Imagine yourself standing on the western wall of Vendelinus or Petavius at the time when this photograph was taken. You would be in a blaze of pitiless, untempered sunshine, but glancing down the precipice at your feet you would seem to be looking into a gulf of blackness. But for the light reflected back from the eastern cliff, and that coming from the earth, there would be scarcely a ray of illumination on the rocks below you. You would look down into inky darkness, and would scarcely dare to make a step from fear of falling over the edge of a bottomless pit. At the same time, as I told you last night, you would see the stars all about you in the sky, even close to the sun.

"This is the reason," I continued, "why the march of day across the moon, always keeping sharp on the heels of night, is a spectacle so imposing and unparalleled. It is this wonderful march that we are going to follow with the aid of the photographs. I shall now ask you to give your attention to photograph No. 3. It was made more than a day and a half later than the others, measured by the age of the moon, which, in this case, was about five days and a half. You notice how in the interval the sunlight has swept eastward over the moon's surface. The *Mare Crisium* is recognizable in the lowest or most northerly, of three large, dark plains. The small white oval a considerable distance above it is our old acquaintance Langrenus, whose floor and walls are now very brilliant in the full sunshine, which falls upon them at a high angle. Vendelinus and Petavius are less conspicuous. The broad, dark plain which has come into view eastward from Langrenus is the *Mare Fœcunditatis*, which we may translate 'Sea of Fecundity'! You certainly cannot aver that on this occasion the invention of the old astronomers failed in the matter of romantic suggestiveness. The name calls up pictures of a great body of tranquil water, fanned by gentle, stimulating breezes, filled with fish of every variety, dotted with vine- and flower-garlanded islets, and bordered by well-watered shores, rich with vegetation, and supporting a numerous and happy population. Some such idea of the *Mare Fœcunditatis* may have been in the minds of its sponsors a couple of centuries ago. But telescopes have become too powerful in our day to permit us to be any longer deceived as to the actual nature of this singular lunar region. Like the *Mare Crisium*, it may have been the bed of a sea many years ago, but at the present time it contains no water, and its shores present an endless succession of fire-scarred cliffs, peaks, and volcanoes. The only 'islands' in it are extinct craters."

"But," said my companion, smiling, "where then is its history?"

"Ah!" I replied, "is not this old sea itself history enough? When it has receded sufficiently into the past, all history loses its details, and presents only its setting and its grand primary elements. Suppose that, some ages in the future, you should be an inhabitant of a distant planet, surveying with a telescope the dried-up basin of the Atlantic Ocean. Provided only that you were convinced, in your own mind, that it had once been an ocean, with fertile, inhabited shores, and with ships sailing upon it, you would be singularly lacking in imagination if you could not reconstruct its history for yourself. The details could safely be left to your invention and you could change them from time to time to suit your varying moods. Terrestrial historians have sometimes done that."

No. 3. July 29, 1903; Moon's Age 5.54 Days.

"But do *you* believe that the *Mare Fœcunditatis* was ever such a sea, and the scene of such events?"

"That is certainly a very pointed question. Questions of that kind are always in order when one is treating of ascertained verifiable facts, but just now, you know, we have wandered a little aside from the straight path of scientific exactitude. Still, I will be frank with you and say that I really possess no settled opinion concerning the former condition of the moon, except so far as what we may call its 'geological' history is revealed by its present state. I am sure that the moon was once the seat of tremendous volcanic action, and I think it not improbable that its great depressed plains were once occupied by water, but as to inhabitants, I know no more about them than you do. Still, I am disposed to think that, as we go on, *you*, at least, will reach the conclusion that all life has not yet disappeared from the moon. We are going to learn some very suggestive and significant things before we are through.

"Farther toward the south and closer toward the terminator you will see in the photograph a third dark plain with five sides, the northern one convex and ill-defined. At its upper corner is an incomplete ring plain. This region bears a still more curious name than the *Mare Fœcunditatis*. It is the *Mare Nectaris* or 'Sea of Nectar.'"

"Apparently your astronomers of old took the moon for an abode of the gods."

"Yes, or for their wine cellar. But we shall get a better look at the surroundings of this Sea of Nectar in a later photograph, and then I shall have more to tell you about it. In the meantime let us return to the *Mare Crisium*. To the east (right-hand side) of the *Mare Crisium* you will observe a diamond-shaped district, not very dark, with a bright point at the corner which faces the *Mare*. You could never guess its name. It is called the *Palus Somnii*, which may be translated 'Marsh of a Dream.' It is a very singular place, and, seen with the telescope, possesses a color which is unique upon the moon, a kind of light brown, quite unlike the hue of any of the other plains or mountain regions. It is covered all over with short, low ridges, as if its surface had been broken up in a most irregular manner with a giant plow. What the person who named it saw there to lead him to connect it in his mind with dreams I have never been able to imagine. The bright point on its western edge is a remarkable crater mountain, named Proclus. What that mountain is made of nobody knows, but it gleams with extraordinary brilliance when the sun strikes it."

"Why may it not be snow-covered?"

"That is a suggestion which has often been made, but one great objection to it is that we have reason for believing that snow, at least in such a situation, cannot exist on the moon. Another objection is that only a few of the lunar mountains are comparable in brightness with Proclus, and they are not the loftiest ones. Upon the whole it is much more probable that the reflecting power of Proclus is due to the composition of its rocks, perhaps to broad crystalline surfaces exposed in the sunshine."

"It is a surprise to me, then, that that 'earthly godfather' of lunar wonders, who had a sufficiently vivid fancy to invent the 'Marsh of a Dream' close by, did not name this mountain for some jewel, real or imaginary."

"It would have been more poetic, indeed, but as I have already told you, the mountains and volcanoes of the moon nearly all bear very prosaic designations, while a wealth of fancy has been lavished in naming the 'seas' and plains. The astronomer Riccioli is responsible for most of the commonplace nomenclature that we find in lunar charts. If you will now glance at the northern (lower) 'horn' of the moon in the photograph you will notice, near the terminator, about two thirds of the way from the *Mare Crisium* to the end of the horn, a pair of ring plains, or crater rings, apparently almost touching one another. They are Atlas and Hercules, the latter being the smaller one on the right. A darker oval below them near the bright edge of the moon is Endymion."

"That, at least," exclaimed my companion, delighted, "is a romantic and appropriate name! I am enchanted to think that Endymion has not been separated by your cold-hearted science from her who loved him so well."

"But if you should look at Endymion with a telescope you would wonder what the moon could find in him to admire. He has been turned into a huge, broken-walled ring plain. You will observe that the other, the southern or upper horn of the moon in the photograph, appears extraordinarily roughened. It is completely pitted with craters and rings. There are so many of them, and they are so entangled, that I shall not undertake to indicate them by their individual names, especially as there is none among them of the very first importance. If, however, you will bring your attention back to the *Mare Nectaris* I shall be able to point out to you a very extraordinary object, which lies just on the border between day and night here, but will be seen in the next photograph that we examine, in full morning light. The object that I mean is a ring on the right-hand edge of the *Mare Nectaris*. Its eastern wall and the top of its central peak are brightly illuminated by the rays of the rising sun; while beyond it, to the eastward, everything, with the exception of the tips of one or two high peaks, is steeped in night. This is one of the mightiest volcanic formations that the moon contains. Its name is Theophilus. To see it and certain gigantic neighbors that it has, fully displayed, we shall turn, after this glance at its first appearance, to photograph No. 4.

"In this photograph the sunrise line on the moon has advanced so much farther eastward that the *Mare Nectaris* lies well within the illuminated part of the disk, and Theophilus has become the most conspicuous object of the kind in view. You now observe that it does not stand alone, but is linked, so to speak, with another similar ring on its southeastern side, while still farther southward is a third less regular ring which seems to belong to the same group."

"Oh, yes," cried my companion, "they certainly do seem to be connected. They look like three links of an enormous broken chain dropped upon the moon."

"The ring nearest to Theophilus," I continued, "and whose northwestern side has been destroyed to give room for the full circle of the wall of Theophilus, is named Cyrillus. The other more distant one is Catharina. If you wish to become a little learned in the geography of the moon it is necessary that you should remember these names. As to the objects that the names designate, they are far too wonderful ever to be forgotten, and it is impossible to confuse them with any other features of the lunar world. There is a great deal of 'history' connected with these three enormous volcanic formations, but I am going to reserve that for a while, because by and by we shall examine a larger photograph of these same objects in which you will see their marvelous details displayed. Now let me direct your attention to the first chain of mountains that we have found upon the moon. Above Catharina you will notice a thin, crinkled line of light passing through a comparatively level district and ending at another ring. It is a range of peaks and cliffs named the Altai Mountains. They are of no great height, and cannot be compared in magnificence with the lunar Alps and the lunar Apennines which we shall see in the photographs taken a few days later, but they are nevertheless very interesting. The ring mountain at which the Altai range terminates is named Piccolomini. It is another marvelous object for telescopic study. The incomplete ring, with a dark interior, which forms the southern corner of the *Mare Nectaris*, resembling a semicircular bay, is Fracastorius. It is a very curious object because close inspection reveals that the missing part of its ring has been submerged, but is still faintly visible through the surface of the *Mare*."

No. 4. November 24, 1903; Moon's Age 5.74 Days.

"I suppose it cannot be water that has covered it, since you have so often assured me that there is no water on the moon."

"No, it is not water, but rock or sand or solidified lava, or some kind of solid matter. It looks as though the whole bed of the *Mare Nectaris* had welled up in one mighty convulsive outpouring of liquid lava, which broke down the wall of Fracastorius, inundated the interior, and then hardened like a floor of cement. The probability that a catastrophe of the kind I have described has occurred here is heightened by the fact that the bed of the *Mare Nectaris* is concave, sunken in the center, as if it had broken and settled down 'like ice upon a pond.' Scattered more or less all over its surface and particularly near its shores, there are indications of this breaking down, and of something that has been covered up."

"To me it seems very mysterious," said my friend, "and very terrible also."

"It is more or less mysterious to the astronomer likewise. Still, geology shows that there have been somewhat similar occurrences on the earth. If you will now direct your eyes to the lower (northern) part of the photograph you will notice some additional things that have come into view with the advance of the sunlight. You observe that a vast somber region occupies the inner portion of the crescent below the center. This consists of two immense plains, one of which sends a large 'bay' as far south as the ring of Theophilus, where it is connected by a narrow 'strait' with the *Mare Nectaris*.

"Turning to photograph No. 5 we see the two plains to which I have referred more fully displayed. The sun has now risen over their entire surface. The upper one is the *Mare Tranquillitatis*, 'Sea of Tranquillity'; and the lower one the *Mare Serenitatis*, 'Sea of Serenity.'"

"I have always thought that astronomers must be happy persons," said my companion, with a smile, "and these names are convincing."

No. 5. July 1, 1903; Moon's Age 6.24 Days.

"Yes, perhaps, but then in bestowing the names they may have been transferring to the moon ideals of tranquillity and serenity which they did not find realized upon the earth. I am not going to talk about these two 'seas' at present because they are better represented upon one of the large photographs which we shall examine later. I prefer to direct your attention just now to some other things. In the first place look once more at Theophilus and its companion rings, and observe how they maintain their preëminence. The entire surface of the moon to the eastward and southward is broken and heaped up with mountains, craters, and rings, but nowhere do we see anything comparable with Theophilus except, perhaps, far toward the south, where near the inner border appear two still larger, but less regular, rings lying in line at a right angle to the terminator. The one on the left is Maurolycus, and the other, still half obscured by night, is Stöfler."

"The names of old astronomers, I suppose."

"Yes, astronomers sufficiently famous in their day, but who would be virtually forgotten at the present time if their friend Riccioli had not thus immortalized them. You see it is a great piece of good fortune to have your name in the moon. It is a kind of revenge for the neglect of future generations at home."

"And it seems to me an equal good fortune to have had an admirer willing to set your name up in the moon."

"Surely. But Riccioli's own name is there also. Afterwards I shall show you his lunar monument, a truly magnificent one. Permit me now to tell you that Maurolycus is much greater in extent than any of the rings that we have yet seen. Not by any means so perfect in form as Theophilus, it covers a vast extent of surface, as much as 150 miles across, with an amazing mass of broken rings, walls, ramparts, ridges and chasms. Some of its peaks are 14,000 or 15,000 feet in height. It has a very lofty central mountain, visible in the photograph, and whose peak comes into view when the sun is rising long before the surroundings have been illuminated, so that it resembles a star glowing amid the blackest night. The neighbor of Maurolycus, Stöfler, is equally extensive and almost equally wild and magnificent when the sunlight is leaping across it from pinnacle to pinnacle and ridge to ridge. In this photograph, however, it is too near the terminator to be well seen. We shall presently pass to photograph No. 6, where Stöfler appears in full light, but before doing so let us glance at the northern part of the moon as here pictured. Close to the terminator, below the grand oval form of the *Mare Serenitatis*, you will perceive two rings, one above the other. They seem to be the complement of the other pair, Atlas and Hercules, which we looked at when the sun had recently risen upon them in another photograph, and which now appear far off toward the west. You observe that Atlas and Hercules lie upon an east and west line, and the others upon a north and south line. The northernmost one is named Aristoteles, and the other Eudoxus. They are situated near the edge of a plain called the *Mare Frigoris*, 'Sea of Cold,' thus named, I suppose, because it lies so far north. Aristoteles is about 60 miles in diameter, and its immense wall is very high and splendidly terraced. Eudoxus, equally deep, is only 40 miles in diameter.

"Turning to photograph No. 6, taken when the moon was more than a day older than it was when No. 5 was made, we have a striking example of the effect of libration in presenting the moon at perceptibly different angles to our line of sight at corresponding phases. We have now arrived at First Quarter, and behold all the western half of the moon illuminated by the sun. You will perceive that we now have in view, simultaneously, six of the great plains called 'seas,' namely, the *Mare Crisium*, the *Mare Fœcunditatis*, the *Mare Nectaris*, the *Mare Tranquillitatis*, the *MareSerenitatis*, and the *Mare Frigoris*, while others are beginning to emerge out of night on the east. Maurolycus and Stöfler, the pair of giant rings in the south, are better seen than before because daylight has advanced farther across them. In fact Stöfler now appears more imposing than its great neighbor, and a smaller ring breaking the continuity of its wall on the western side is visible. Above these, in the direction of the south pole of the moon, and around the pole itself, the surface is marvelously rough and broken. It looks as if it would be impossible to find a level acre of ground in all that region. The rings and craters are veritably innumerable. It is the existence of these irregularities which causes the terminator to appear so crooked and broken. At some places you perceive small bright points within the edge of the night half of the moon. These, of course, are the summits of peaks, which have just been touched by the sunlight while the surface all around them is still covered with darkness.

"Below Stöfler, all along the terminator, as far as the middle of the moon, an irregular row of rings appears. Three of these bear some resemblance to the great group of which Theophilus is the chief member. They are, counting from south toward north, Aliacensis, Werner, and Blanchinus. Below them two other much larger ones are conspicuous, Albategnius, the more southerly, and Hipparchus. These two are full of moon history. Albategnius, the smaller, is very deep and comparatively perfect in condition, while Hipparchus, more than 90 miles across, has been vividly described as a 'wreck and ruin,' its walls, once possibly of great height, being now low and broken, and traversed with gaps and valleys, while a great cleft exists crossing a part of the broad, irregular floor. It is probable that Hipparchus is an older formation than Albategnius."

No. 6. November 26, 1903; Moon's Age 7.75 Days.

"Pardon me," interrupted my companion, "but I must cry for mercy. Really, these strange names escape from my mind as fast as you mention them. Is there not something a little more romantic in the moon—something to relieve the strain of all this nomenclature of words terminating in 'us,' and this frightful lunar geology?"

"Yes," I said, "I believe that on the other half of the moon, which has not yet seen the sun rise, we shall find something better to your taste. But do not be too impatient. Reflect that these names represent very wonderful things visible to us in another world than ours, things the knowledge of which has cost the lifelong labors of many gifted men, and that will be remembered, studied, talked, and written about centuries after we are dead. Fortunately for your powers of attention the eastern half of the moon, upon which day will be seen gradually dawning in the next set of photographs, has a general character quite different from that of the western half. It contains the greatest ranges of lunar mountains, yet upon the whole it is more level, being covered to a great extent with broad plains, in the midst and along the borders of which stand the most remarkable and interesting of all the lunar formations. In and around some of them we shall search for the evidences which some astronomers think that they have found of life upon the moon."

"Oh, that indeed will be interesting!" exclaimed my friend with reviving animation.

"But," I added, "do not place your expectations too high. Keep your imagination under control, try always to be just a little 'scientific' in your way of looking at things, and then I believe you will not be disappointed."

"Oh, please do not think that I have been disappointed," she said deprecatingly. "But positively you must admit that 'Albategnius,' 'Aliacensis,' 'Blanchinus,' and 'Maurolycus,' are not precisely captivating. Remember that I have read little except poetry and romance, and those histories that are full of stories."

"You will find a deep vein of poetry and romance in the moon," I replied, "before we have finished, and after you have reflected upon what we have seen and what we have been saying."

Leaving the remaining photographs to be examined after lunch, we now entered the house.

II FIRST QUARTER TO FULL MOON

II FIRST QUARTER TO FULL MOON

NOTWITHSTANDING the signs of impatience which my friend had manifested when we were passing, in our review of the photographs, from one lunar ring mountain to another, all more or less similar in appearance and characteristics, I was gratified to see that her mind was still attracted to the subject of the moon, and during the lunch she, of her own accord, began to talk of it.

"You have said so much about volcanic occurrences on the moon," she remarked, "that I wonder why you do not call those immense mountains 'volcanoes.' I observe that you always speak of them as 'rings,' or 'mountain rings,' or 'ring plains'; while to me, although to be sure I am no geologist and have perhaps no right to an opinion, they seem plainly to be just huge volcanoes and nothing else."

"Your observation is quite correct," I replied, "as far as superficial appearance goes, and I may add that these great rings are often called volcanoes. If we apply the proper adjective and name them 'lunar volcanoes,' perhaps there can be no objection to the term. But they are certainly widely different from our terrestrial volcanoes. The difference is not in size alone, although in that regard it is enormous. There is a far more significant difference, which you could hardly be expected to notice in a simple inspection of the photographs, although it is evident when once pointed out. I refer to the fact that what seem to be the craters of lunar volcanoes are not situated on the tops of mountains. They are immense plains, more or less irregular in surface, and often having a peak or a group of peaks in the center, while around these plains always extends a mountain ring, steep on the inner side, and having a gradual slope without. But most significant fact of all, the plains, or floors inside the ring, are almost invariably situated thousands of feet below the general level of the moon. If the terrestrial volcanoes were formed on the plan of the lunar ones, when we visit Vesuvius, instead of climbing up a mountain rising out of the midst of a plain and capped with a cone, having a funnel-shaped crater in the center, we should find before us a relatively low, circular elevation, on surmounting which there would appear on the inside of the circle a great basinlike hollow, far below the level of the surrounding country. In the center of this, distant from the lofty encircling walls, would be seen a conical hill with smoke and vapor issuing from a vent at its summit. The top of this crater hill would be lower than the rim of the basin-shaped hollow, so that the whole volcano with its immediate surroundings would be inclosed and shut off from the environing upper world by the sides of the basin. While you finish your coffee I will make a sketch which may render this difference between lunar and terrestrial volcanoes evident at a glance."

Lunar Volcano, in Section.

Terrestrial Volcano, in Section.

Accordingly, after a few minutes, I presented to her these two diagrams, remarking that it should be borne in mind that the two sketches were not made on the same relative scale. "I was compelled," I said, "to change the true proportions in the section of the lunar volcano, for if I had drawn them as they are in fact, the width of the basin would have been enormous in proportion to its depth. You will recall that I told you that such rings as Albategnius and Maurolycus are a hundred miles and even more in diameter, while their depth does not exceed two or three miles. It results from this necessary falsification of proportions in the sketch that the terrestrial volcano, although so widely different in form, appears comparable in magnitude with the lunar one. But the fact is that you could take a dozen of the largest volcanic mountains on the earth and throw them into one of the great lunar rings without filling it."

"I am the more astonished by what you say," remarked my friend, "because you have already told me that the moon is so much smaller than the earth. How does it happen, then, that her volcanoes are so much larger? I should think that in a little world all things would be small in proportion."

"It is quite natural to think so," I replied, "until you reflect upon the consequences of the smaller force of gravitation on a small world. I told you last evening that gravitation on the moon, is only one sixth as powerful as it is on the earth, and you will recall that one consequence which I pointed out was that you would weigh only twenty pounds if you were on the moon.

Since the same reasoning applies to all objects in the lunar world, it is clear that a similar force exerted there would be able to produce enormously greater effects, as for instance in the formation of vast hollows or depressions, by violent explosions, the products of which would be thrown to immense distances. Some selenographers, which is a term applied to those who study the features of the lunar world, have suggested that in this cause alone is to be found the explanation of the giant lunar ring mountains. At some remote period of the past, according to them, the volcanic forces of the moon reached a maximum of activity and energy. The lava, cinders, ashes, and other products of ejection, were hurled to a height of scores of miles, and when this fell back at a great distance from the centers of eruption these were piled up in huge rings, fifty, eighty, or a hundred miles in diameter, while the surface of the moon within the rings sank in consequence of the withdrawal of the material thus ejected. To account for the existence of the central mountains so often found in the middle of the rings, it has been suggested that at a much later period, when the volcanic energy had become comparatively insignificant, as a result of the cooling of the interior of the moon, less violent explosions, not greater than many that have occurred on the earth, took place, and by these the central peaks were formed."

"You are going to think me too romantic, or too imaginative, again," said my friend, with a smile, "but I cannot prevent myself from wondering what the inhabitants of the moon did and thought while all those marvelous things were happening."

"I have not said that there were inhabitants of the moon."

"No, but you have confessed that there might have been inhabitants, some time, and I should like to know whether they were there when those terrible volcanoes were formed."

"If they were," I replied, "they could not have survived such a universal upheaval as the surface of the moon has undergone. You have seen in the photographs that the great rings and smaller craters are scattered thickly over the moon. It is true that comparatively few are found in the level expanses called 'seas,' but if those regions were covered with water they could only have been inhabited by beings provided with gills and fins."

"How long ago did these explosions occur?"

"I cannot tell you, except that it must have been many ages in the past; so long ago, indeed, that the whole course of human history seems but a day in comparison."

"Then," said my friend with animation, "there has been time enough *since* that dreadful period for inhabitants to develop upon the moon, has there not?"

"Yes, time enough, perhaps, provided that sufficient water and air and other vital requisites remained after the exhaustion of the volcanic energies."

"Oh, let us say that they did remain. I am eager to believe that the moon has not always been so desolate as she appears at present."

"Very well, you are at liberty to believe that if you like. No astronomer is likely positively to contradict you, although he may smile a little incredulously. Besides, as I have already told you, there are certain rather inconclusive indications of some kind of life, and of some kind of activity, still on the moon."

"Please show them to me, then, or tell me about them. Perhaps I shall find them less inconclusive than you do."

"Everything in its turn," I replied. "We shall come to the indications that I have spoken of after we resume the inspection of the photographs."

"Then I am ready to resume at once."

Accordingly we returned to the table and the photographs under the pleasant shade of the elm. Taking up the photograph numbered 7, I remarked that it exhibited the moon as it appears a little after First Quarter; that is to say, a trifle more than half the face turned toward the earth is in the sunlight. I called attention once more to the six "seas," which we had already remarked, and to the continued conspicuousness of Theophilus and its companions, a little above the middle of the visible hemisphere.

"You observe now," I continued, "how the rotundity of the lunar globe begins to manifest itself as the sunlight sweeps farther eastward. The crescent shape is gone and the line between day and night begins to be bowed outward, convexly. The *Mare Crisium* is particularly well defined, and also the diamond-shaped region called the *Palus Somnii*. With the sun so nearly vertical above it, the remarkable peak of Proclus, between the *Palus Somnii* and the *Mare Crisium*, has become very brilliant. In a telescope you would see it glowing almost like a star. You observe also that several long, straight, bright rays proceed from it in several directions."

"All the more reason, it seems to me," said my friend, "why your unimaginative astronomer, Riccioli, should have named it for some brilliant gem instead of attaching to so dazzling an object the prosaic designation of 'Proclus.'"

"After all," I replied, "what's in a name?" Now that you are familiar with the appearance of Proclus, its name will henceforth call up to your mind an image as brilliant as if it had been named 'Mount Diamond' or 'Mount Amethyst.'"

No. 7. July 2, 1903; Moon's Age 7.24 Days.

"Pardon me," said my friend, "but it was not of names like those that I was thinking. Observe how he who named the neighboring *Palus Somnii*, 'Marsh of a Dream,' exhibited an exquisite delicacy of fancy. It suggests something indefinitely strange, romantic, imaginative. That unknown astronomer, unknown at least to me, put a little of himself, a little of his inmost mind, into the name, and I thank him for it. I shall never forget the 'Marsh of a Dream' in the moon. It will haunt my own dreams. I shall be all my life seeking and never finding its meaning."

"Since you are in so poetic a mood," I responded, "I rejoice that besides its bald facts, its fireless volcanoes, and its dried-up plains, the moon possesses many things that can stir the imagination of the most sentimental observer. But, in order that we may not wander too far from the paths of science, let me recall your attention to the photograph. We have been going over ground already trodden by returning to the neighborhood of the *Mare Crisium*. I shall now lead you back to the terminator, where we shall find a little that is new. Still nearly hidden in night we perceive many great rings on which the sun is beginning to rise, and four of the most important ranges of mountains are coming into view. One of these, on the southern border of the *Mare Serenitatis*, is visible throughout its entire extent. It forms a portion of the coquettish ornaments with which the Moon Maiden has decorated her hair, as we shall see clearly in the next photograph. This range is named the Hæmus mountains. Near its center, quite at the edge of the 'sea,' is a bright crater ring, one of the most conspicuous on the moon. It is called Menelaus."

"Menelaus?" exclaimed my friend. "Ah, then Riccioli did not confine his favoritism to the astronomers and philosophers in putting their names in the moon. Menelaus, if I remember my classical reading correctly, was the husband of Helen of Troy."

"Yes, the brother of Agamemnon himself. You must admit that Riccioli occasionally felt his imagination a little awakened. He was not altogether destitute of the spirit of poetry."

"But did he also put Helen in the moon?"

"I am sorry to say that he did not. It would have been a very suitable abode for her. However, if you like, you may recognize Helen in the Moon Maiden herself."

"Thank you, that will be, indeed, an unexpected pleasure."

"Meanwhile allow me to point out to you that there is a curious light streak, very faintly shown in the photograph, which crosses the *Mare Serenitatis* from Menelaus to the opposite shore, and reappears more distinctly, on the lighter-colored plain toward the north. This streak comes all the way from a great ring mountain named Tycho in the southern part of the moon. It is more than 2,000 miles long, and is one of the greatest mysteries of the lunar world. Tycho, which lies just on the sunrise line, is not well seen in this photograph. It has a great number of these strange streaks or rays proceeding from it in all directions. We shall study them in one of the photographs which are to come. One word in regard to the plain north of the *Mare Serenitatis* of which I have just spoken. It, too, has a name that is calculated to appeal to your lively imagination. It is called the *Lacus Somniorum*, which if my knowledge of Latin is correct, means 'Lake of the Sleepers.'"

"Then your old friend Riccioli certainly did not bestow the appellation."

"No, it was one of his more fanciful, or, if you prefer, more poetical predecessors, perhaps the same who imagined the 'Marsh of a Dream.'"

"Oh, that gives me another reason to think of him with admiration and gratitude. He, at least, had a soul that rose above mere prosaic facts."

"Perhaps. But do not think too lightly of the facts of the moon. After all the human mind must base itself upon the solid ground of fact. Without that we should become mere dreamers, and be suited only to inhabit your favorite 'Marsh.'"

"The other mountain ranges of which I have spoken," I continued, "are faintly distinguishable eastward from the *Mare Serenitatis*. They are the Apennines, the Caucasus, and the Alps. But perhaps we had better turn at once to photograph No. 8 where they are much more clearly seen, because the sunrise there has advanced a couple of hundred miles farther east."

"But, dear me, how slowly the sun rises on the moon! Was this photograph taken a day later than the other?"

No. 8. August 31, 1903; Moon's Age 9.22 Days.

"Almost exactly two days later. When it was made the moon was nearly nine and a quarter days old, and its age at the time No. 7 was made was only seven and a quarter days. But, owing to the effects of libration, an explanation of which I have put into a note for your private reading when you feel like it, [see p. 57, footnote], the difference of phase amounts to less than two days. You are right, however, in remarking that sunrise is a very slow process on the moon. It requires about two weeks to pass from the western side of the moon to the eastern side, and both day and night at any point on the moon last about a fortnight. This results from the fact that, as I have told you, the moon does not turn rapidly on its axis like our own globe, but keeps always the same side directed toward the earth. Accordingly, a lunar day and night are together about a month long."

"And was it so when, as I must persist in believing, there were inhabitants on the moon?"

"Probably, although it may have been shorter then. The consequences of these excessively long days and nights would be very serious to beings fashioned upon the terrestrial plan. In the practical absence of an atmosphere the heat of the sun's rays, pouring down without interruption and without the intervention of any clouds or vapors for fourteen days at a time, must be simply overpowering. And then, during the equally long night that ensues, the radiation into open space must quickly leave the surface of the moon exposed to the most frightful degree of cold, comparable with the absolute zero of empty space!"

"But think, what a merciless environment you are picturing for my inhabitants of the moon. Please do not forget that I insist that their comfort shall be considered."

"Oh, as for that, you know you were content a little while ago to relegate your inhabitants to a remote period in the past, after the volcanic fury of the lunar world had ceased, and before its present airless and waterless condition had supervened. Possibly at that time things were not so uncomfortable for them. They may have had clouds to temper the sunshine, rains to cool the days and dews the nights, and shady parks like yours for philosophic and scientific contemplation."

"Do not forget the poets."

"Certainly not. But is not the moon herself the very spirit of poetry? What in nature is more poetical in its suggestions than the moon wading through fleecy clouds on a serene summer's night? But pardon me, we are forgetting my mountains, upon which I insist as strongly as you do upon your inhabitants. The mountains have this advantage that they are very real, and no exercise of the imagination is required to bring them clearly before us. In photograph No. 8 they are all visible. The Apennines, the greatest of them, start from the eastern end of the *Mare Serenitatis*, and run in a slightly curved line southeastward, a distance of about 450 miles. They form the singular ornament which the Moon Maiden (or shall we now call her Helen of Troy?) wears upon her forehead. Turn the photograph upside down so that the moon is presented as the naked eye sees it in the sky, and you will find that, although he aimed only to be scientifically exact and to exclude everything but the real facts, Mr. Wallace has produced an excellent picture of this wonderful face in the moon."

"But what is that face?"

"It is humanity projected upon the moon. It is a lesson on the powers of the imagination. We perceive a certain collocation of mountains, peaks, and plains on the disk of the moon, and our fancy sees in them a human likeness. We should congratulate ourselves that we are able to do this. It is a kind of proof of superiority. Many brute animals do not recognize even their own likenesses in a mirror, much less in a picture. But the Moon Maiden is perhaps as real as your inhabitants."

"I am not prepared to confess that yet."

"Very well, let us go on. The lunar Caucasus is the broader, but shorter, range of mountains at the northeastern corner of the *Mare Serenitatis*, and the Alps extend eastward from the Caucasus to a conspicuous dark oval close to the terminator, which is one of the most remarkable formations on the moon, and which, when we come to study it in one of the larger photographs, will probably interest you deeply because it is one of the places where recent studies have discovered indications of what may possibly be some form of lunar life. I wish now to direct your attention to the central and upper parts of the photograph. Running downward from the south, a little west of the terminator, you will perceive a double row of immense rings and ring plains. They are not only remarkable individually, but quite as remarkable for their juxtaposition in two long ranges. Among them, in the westernmost row, are three or four whose names you may remember—Maurolycus, Stöfler, Aliacensis and Werner. Still larger ones are included in the eastern row, the largest of all being at the bottom. It is rather a hexagon than a circle. It is 115 miles in diameter, and the flat plain inside the bordering mountains contains about 9,000 square miles. By close inspection you will perceive a small crater mountain near the northwestern side. This immense walled plain is named Ptolemæus after a great astronomer of antiquity, the author of the Ptolemæic system, which treated the earth as the center of the universe.

"Still more interesting are the things visible farther south. You cannot fail to remark a very beautiful ring, a perfect circle, brightly illuminated on the eastern side, and having a bright point symmetrically placed in the exact center. It is named Tycho, after another great astronomer, and is generally regarded as the most perfect crater ring on the moon. It is 54 miles in diameter, and its walls are about 17,000 feet high on the inner side, more than a thousand feet higher than Mt. Blanc, the giant of the terrestrial Alps. Its central mountain is 5,000 feet high. The most remarkable thing about Tycho is the vast system of 'rays' or bands which seem to shoot out from it in all directions, traversing the surface of the moon, north, south, east, and west for hundreds of miles, and never turning aside on account of any obstacle. They lie straight across mountains, valleys, and plains. We have already seen one of them, the largest of all perhaps, crossing the *Mare Serenitatis* and the *Lacus Somniorum*, in the northern hemisphere of the moon. Nobody knows exactly what these rays mean or what they consist of. We shall from this time on see them in all the photographs that we examine, and later I shall have more to say about them, and the speculations to which they have given rise.

"About half way between Tycho and the south pole of the moon, you will see an enormous irregular plain, with lofty broken walls, interrupted by a number of crater rings. Several similar rings also appear in the interior of the plain. If Tycho is the most perfect in form of the lunar crater rings, this great inclosure, which is named Clavius, is the finest example of the walled valleys. It is more than 140 miles across, and covers an area of not less than 16,000 square miles. Two of the rings within it, which seem so small in comparison, are 25 miles across. A smaller walled plain, yet one of really immense size, is seen half way between Tycho and Clavius, and farther from the terminator than either of them. This is Maginus, and it possesses the peculiarity that at full moon it practically disappears!"

"But how can that be possible? I see nothing behind which it can be hidden."

"It is the sunlight that hides it. You must have noticed already that the rings and mountains are best seen when at no great distance from the terminator, because there the sunlight strikes across them at a low angle, and their shadows are thrown sharply upon the adjoining slopes and levels. Look at the western part of the moon in the photograph before us. Many of the huge rings and walled plains that were so striking in appearance when the sun was rising upon them are now barely visible. Langrenus and Petavius, for instance, have become no more than whitish blotches, and even Theophilus is no longer conspicuous. The reason is because when the sunlight falls vertically upon any part of the moon there are no shadows there, and without shadows there can be no appearance of relief. Then the mightiest mountains are almost lost from sight in the universal glare. The same thing would be apparent if you were suspended above the earth at a great height in a balloon and looking down upon the tops of the snowclad Rockies. Without shadows serving to reveal their true character and to throw their outlines in silhouette upon the adjacent plains, they would resemble only white spots and lines on the generally darker expanse of the continent. But Maginus is an extreme case. Owing to the relatively small elevation of its walls, and their broken-up state, and owing also, probably, to a similarity of color between the mountain ring and the inclosed plain, when the light is vertical upon them, as at the time of Full Moon, they blend together and become barely distinguishable from one another, and from the surrounding surface of the moon.

"Take now photograph No. 9. The age of the moon here is actually less than it was in the photograph that we last examined, yet, in consequence of libration, which has caused the moon, in effect, to roll a little to one side, the sunlight is farther advanced toward the east, and we see many features of the lunar world that before had not yet emerged from night. Clavius you will notice is much more fully illuminated. See how distinctly the shadow of its vast western wall is cast upon the floor of the valley within, while the opposite eastern wall with its immense cliffs and precipices glows in full sunshine, its shadow, thrown toward the east, blending with the darkness of night still covering that side of the moon. Southeast of Tycho, which is beautifully shown here, two other great walled plains have come into view. The uppermost of these is Longomontanus and the other Wilhelm I. For a considerable distance below these (toward the north) the surface continues broken with rings and craters, but at length these give place to a dark, level expanse. This is a part of the *Mare Nubium*, or 'Sea of Clouds.'"

"Not quite so romantic a name as some of the others," remarked my friend, "but still I think I can be sure that Riccioli had nothing to do with the selection. There is certainly something poetic in the idea of a sea of clouds."

"It is a very beautiful region when examined with a telescope," I continued, "and its mountainous shores contain many interesting formations. Farther north, you will observe, near the terminator, and apparently lying in the midst of the *Mare Nubium*, a large ring, as perfect in form as Tycho itself. This is a very famous object, and it bears the name of the great astronomer Copernicus, who overthrew the Ptolemæic system and established in its place the true idea of the solar system, namely, that the sun is its center, while the earth and the other planets revolve as satellites around him."

"Surely," said my friend, "Copernicus deserved to have his name placed in the moon, and very conspicuously, too."

No. 9. August 2, 1903; Moon's Age 8.97 Days.

"It could not have been made more conspicuous," I replied, "for the situation of the great ring mountain called Copernicus, in the midst of an immense level expanse, makes it one of the most marked features of the lunar world. Copernicus is the subject of one of the larger photographs that we are going to examine later, and I reserve a description of its peculiarities. North of Copernicus you will observe apparently a continuation of the *Mare Nubium*. But it is really another 'sea' that we are looking upon there, the *Mare Imbrium*, 'Sea of Rains.' The baylike projection that runs out into the bright highlands west of Copernicus bears the name of the *Sinus Medii*, 'Central Gulf,' and the one just below it is the *Sinus Æstuum*, 'Gulf of Heats,' which is certainly suggestive of dog days on the moon. Observe that the *Sinus Æstuum* merges on the west with a dark, oval area, which is called the *Mare Vaporum*, 'Sea of Mists.' It is one of the darkest districts on the moon. If you will now turn the photograph upside down you will find that the *Sinus Medii* constitutes the dark eye of the Moon Maiden, while the *Sinus Æstuum* and the *Mare Vaporum* form that portion of her hair which droops upon her forehead."

"Why not frankly call it frizzed?"

"Because I feared that you would not consider that a sufficiently poetic term."

"But I find poetry enough in the names 'Gulf of Heats' and 'Sea of Mists.' My admiration for the man who could think of such appellations continually increases."

"Then please reverse the photograph, for we must not lose ourselves in dreams. You will notice that the range of the lunar Apennines runs between the *Mare Vaporum* and the *Sinus Æstuum* on one side, and the *Mare Imbrium* on the other. The entire chain of the Apennines is beautifully shown here. They are exceedingly steep on the side facing the *Mare Imbrium*, and gigantic peaks standing upon their long wall cast immense shadows over the 'sea.' Their southwestern slopes are comparatively gentle, rising gradually from the level of the *Mare Vaporum*. At their upper or southern end, in the direction of Copernicus, they suddenly terminate with a beautiful ring, which is called Eratosthenes. This is a fine example of the disk or cup shape of the lunar volcano. The bottom of Eratosthenes lies 8,000 feet below the level of the surrounding *Mare*, while peaks on its wall are as much as 15,000 or 16,000 feet in height. Between the lower end of the Apennines and the upper end of the Caucasus Mountains a strait opens a broad, level way between the *Mare Imbrium* and the *Mare Serenitatis*. On one of the large photographs these two 'seas' and the strait connecting them are represented in all their picturesque details, as you will see when we come to study them. I promise you at that time a free rein to your imagination and plenty of room for its flights. On the northern border of the *Mare Imbrium* and close to the terminator we see once more the remarkable oval valley to which I referred when pointing out the lunar Alps, and which bears the name of Plato. I call your attention to it and also, again, to Copernicus, in order that you may compare their appearance here with that which they present in the next photograph, taken when the moon's age was eleven and three-quarter days."

No. 10. November 30, 1903; Moon's Age 11.78 Days.

We hereupon turned to photograph No. 10.

"Now," I continued, "observe the difference that some two days' advance of the sunlight has produced. Plato is far within the illuminated part of the disk, and it looks darker than before. Copernicus, on the other hand, which appeared as a sharp ring with one border dark when it was near the sunrise line, has now become a round, white spot, somewhat darker in the center, with a great grayish splatter surrounding it upon the surface of the *Mare*. In the meantime, over nearly the whole extent of the *Mare Imbrium* the sun has risen and two other *mares* have made their appearance, one of which, extending across half the width of the eastern hemisphere, might be called the Pacific Ocean of the moon, if it had any water. It is named the *Oceanus Procellarum*, the 'Ocean of Tempests,' while at its southern extremity a very dark nearly circular expanse, inclosed with mountains, bears the name of the *Mare Humorum*, 'Sea of Humors.'"

"Evidently the astronomer who bestowed that name was not in a joking mood else he would surely have called it the 'Sea of Humor.'"

"No, apparently he was in deep earnest. But what kind of humors he was thinking of I cannot tell. Perhaps the name occurred to him because the *Mare Humorum* is the darkest of all the great levels on the moon. It is very conspicuous to the naked eye at Full Moon. You will perceive that

Tycho has now become the most prominent of all the rings on the moon. It will maintain this distinction and continue to gain in conspicuousness up to the time of Full Moon. Seen as we now see it, Tycho manifestly merits the appellation sometimes bestowed upon it of the 'metropolitan crater of the moon.' Notice how bright the mysterious bands radiating from it have become. The higher the sun rises upon them the more brilliantly they glow, almost as if they were streaks of new-fallen snow. They spread over the whole of the southwestern quarter of the moon, hiding rings and mountains with their brightness. One very notable ray runs down into the *Mare Nubium*, and a fainter one parallel with it produces the semblance of a long, walled way.

"The South Pole of the moon lies in the midst of a marvelously upheaved and tumbled region, where one huge ring is seen breaking into another on every hand. One of these rings, named Newton—it lies just on the upper edge of the disk, south of Clavius—surrounds the deepest known depression on the moon. Its bottom sinks to a depth of 24,000 feet below the highest point on the wall. This gigantic hole is so profound that, situated where it is, close to the pole, where the sun can never rise very high, its depths remain forever buried in night. It is the very ideal of a dungeon, for if you were imprisoned at the bottom you would never see either the sun or the earth."

"You make me shudder! Truly, after all, the moon appears to be a world filled with dreadful things. Who would ever imagine it, seeing how serene and beautiful she is in a calm night?"

"Yet is there not a kind of beauty even in those things, like the abyss of Newton, which appall you only when you know the real facts about them? There is a certain grace in their shapes and outlines, and a great attraction for the eye in their contrasts of light and shadow. It is the same sort of attraction which we find in such terrestrial scenes as the Yosemite Valley viewed from Inspiration Point, or the awful depths and chasms of the Grand Cañon of the Colorado. The presence of man and his works is not always essential in order to fix our attention upon the wonders of nature. Their very grandeur exalts us until we forget our little race and its ephemeral achievements."

"Still, I hope that you will show me something on the moon less awe-inspiring and suited to awaken more quiet thoughts, and especially to reassure me concerning my lunarians, as I suppose you would call them."

"You shall not be kept long in expectation. Turn your eyes once more to the *Mare Imbrium*. You will observe that its northern shore consists of a series of curves, each terminating with a promontory projecting into the sea. When looking at it I am often reminded of an entrancing view which I once enjoyed from the summit of Mt. Etna over the island of Sicily. From that great elevation nearly the whole eastern and southeastern coast of the island was visible as upon a map. The indented shore stretched away in long, graceful curves, where the blue Mediterranean contrasted sharply with the yellow sands, and the eye, wandering from Catania to Syracuse, was enchanted with the beauty of those geometric lines. But the winding coast of the *Mare Imbrium* is far longer than the shores of Sicily, and the mountains and cliffs bordering it are more wonderful than any corresponding scenes on the earth. I wish, particularly, to have you look at the easternmost of the indentations on the northern side of the *mare*. It bears a designation that must surely please your imagination. It is the *Sinus Iridum*, 'Gulf or Bay of Rainbows.'"

"I recognize the work of my old friend the unknown astronomer. Verily he had a poetic soul! And he has written his poem on the chart of the moon, for those to read who can."

"It is a charming landscape that the telescope reveals there," I said, "even though no rainbows are visible."

"But you will not deny that they may once have spanned that bay and its shores with their exquisite arches?"

"No, I shall not deny so pleasing a possibility. I will only say that it lies beyond the ken, and even outside the field, of science."

"Then I regard it as fortunate that *he* was not too exclusive in his devotion to science, for then he could never have seen the rainbows with the eye of fancy, and your charts would not have been adorned with so delightful a name."

"Let me tell you about this bay or gulf," I said, tapping the photograph to recall her from her reverie. "You observe that it terminates at each end with a promontory. That at the western end is named Laplace, and the other Heraclides. The latter is the more picturesque. If ever you have an opportunity to see the moon with a good telescope do not fail to look at the promontory of Heraclides, for if you are fortunate in the choice of the time of observation when the setting sun is throwing its shadow over the adjoining 'bay,' you will find that the serrated outlines of the promontory represent, in a very striking manner, the profile of a woman, more sharply defined than the face of our familiar Moon Maiden, but a mere miniature in relative size. The shores of the *Sinus Iridum* are bordered with high cliffs, behind which rise the peaks of a mighty mountain mass. Just back of the center of the great bowed shore of the 'bay' appears, in the photograph, a small, bright crater ring. This bears the name of Bianchini. It is a lunar volcano, 18 miles in diameter, rising out of the midst of many ranges of nearly parallel hills and mountains, the general direction of which corresponds with that of the shore of the 'bay.' If there is any place on the moon where one is tempted to think that the scenes of a living world might once have been witnessed it is the *Sinus Iridum* and its neighborhood. Its latitude is between 40° and 50° north, corresponding with the most thickly populated zone of our own globe. The surface of the 'bay'—once its bottom, if we admit that it was ever filled with water—is gently undulating, with winding ridges that suggest the action of tides and currents in sweeping to and fro deposits of sand and gravel, and piling them in long rows of bars and shallows. One can hardly help picturing in the mind's eye waves breaking on the curving beach and dashing against the projecting rocks of the promontories; a white city seated just at the center of the shore of the 'bay,' near Bianchini, like Naples at the feet of Vesuvius; a rich vegetation covering the slopes of the mountain valleys, and romantic sails dotting the 'bay' and the neighboring 'sea.'"

"I am very glad to observe," interrupted my friend, "that you are not hopelessly prejudiced against my opinion that the moon has not always been 'dead,' as you call it."

"I am so far from it," I replied, "that I am half disposed to admit that she is not altogether dead even yet. But it is my duty to keep you as close as possible to the known facts. We shall see the *Mare Imbrium* and the neighborhood of the 'Bay of Rainbows' again. Meanwhile, suppose we turn to the next photograph of the series, No. 11. The age of the moon here is about thirteen days. She is fast approaching the phase of Full Moon. The first thing to which I would direct your attention now is the exceedingly brilliant point of light which has come into view near the terminator, a little north of east where the *Mare Imbrium* merges into the *Oceanus Procellarum*. In several ways this is the most noteworthy object on the moon. It led the famous English astronomer, Sir William Herschel, to believe that he had seen an active volcano on our satellite. He naïvely wrote in his notebook on a certain occasion: 'The volcano glows more brightly to-night!' Yet it is no more active than the other craters and crater rings in the lunar world. It is only extraordinarily, almost incredibly brilliant—by far the most dazzling point on the moon. It is a ring mountain, and is named Aristarchus. It has a near neighbor, barely visible in this photograph, close by toward the east named Herodotus. Herodotus is by no means remarkable for brilliancy. The central peak and a part of the floor and the east wall of Aristarchus consist of some material—nobody can tell what it is—which gleams in the sunlight, I had almost said like diamonds, although that would be an exaggeration. There are three or four other crater rings on the moon, including Proclus, which are also very brilliant, but not one of them can be regarded as a rival of Aristarchus. Its power of reflection is so great that it is even visible with a telescope in the lunar night, when the only light of any consequence that reaches it is that sent from the earth. It was, indeed, this fact which misled Herschel. He saw Aristarchus shining on the night side of the moon, and naturally thought that only the fires of an active volcano could have rendered it thus visible."

No. 11. December 1, 1903; Moon's Age 12.98 Days.

"And are you sure that he was mistaken?"

"Positively. There is no fire in Aristarchus, and has been none for ages."

"But why do not astronomers undertake to find out what it is that makes Aristarchus so brilliant, then?"

"They have almost no data to go upon. You should be informed that even the greatest telescopes, with their highest powers, are unable to bring the moon within less than an apparent distance of say forty miles. At such a distance it is manifestly impossible to tell of what a lunar formation consists. We cannot analyze the moon with the spectroscope as we can the sun and the stars, because she does not shine with her own inherent light. We can only infer that a large part of the substance of Aristarchus consists of something which reflects a very great proportion of the light that falls upon it. If a mountain on the earth were composed of a vast mass of crystals, or of bare polished metal, we might expect it to present, when seen from the moon, some such appearance as we notice when we look at Aristarchus.

"In this photograph the *Sinus Iridum*, having the sun higher above it, is more brilliantly illuminated than in No. 10. Particularly you will notice the brightness of the line of cliffs along its eastern curve, terminating at the promontory of Heraclides."

"That is the promontory which presents the profile of a woman's face, if I recall correctly what you told me."

"Yes. Please observe also that the oval of Plato is as dark as ever, while Copernicus has, if possible, increased in brightness, and the great splatter of broken rays around it seems to have extended farther over the surrounding maria. Almost directly east of Copernicus, in the *Oceanus Procellarum*, appears a much smaller crater ring, Kepler, which resembles a miniature of Copernicus because it, too, is encircled with a kind of corona of short, bright rays. Copernicus, Kepler, and Aristarchus mark the corners of a large triangle. Speaking of rays recalls us to Tycho. You will see that, as I told you, this wonderful formation grows in relative prominence when the period of Full Moon approaches. Its ringed wall and central mountain are obscured by their own brilliance, while the gigantic system of bright bands, or rays, which have their center of origin at Tycho, is gradually becoming the master feature of the bright part of the moon.

"I have told you that the *Mare Humorum*, which is very sharply defined in the picture before us, is the darkest of all the level areas that go under the name of 'seas.' It is not, however, the darkest *spot* on the moon. There are several places where the surface appears, at times, much duskier than in any part of the *Mare Humorum*. Three or four of these are clearly discernible in this photograph. They lie westward from Copernicus in the *Sinus Medii*, the *Sinus Æstuum*, and the *Mare Vaporum*. Their dusky hue strikes the eye at once. They give the impression of sink holes. No special name is attached to them, but they must have been evident to the first observers, with the smallest telescopes, and it is rather surprising they should have escaped special designation on the lunar charts. A fact which will especially interest you is that some observers look upon these and other dusky areas on the moon as being, possibly, indications of the existence of some kind of vegetation there."

"But if there is vegetation there may be other kinds of life also, may there not?"

"Ah, I have not said positively that there *is* vegetation, but *if* there is then your conclusion as to other life may be correct. Glance next at the upper part of the disk along the terminator. Two or three broad oval rings have come into view there. The largest of these with its long eastern wall lying exactly on the line between day and night is an extremely interesting formation, bearing the name of Schickard. The plain within the ring is almost large enough to have been called a 'sea' or at least a 'lake.' It is about 134 miles in diameter, and is in reality much more nearly circular than it appears to be. Like all similar formations situated near the 'limb' of the moon, by which we mean the edge as viewed from the earth, it is greatly foreshortened by perspective. The scale of the photograph is, unfortunately, not large enough to reveal an unique thing in the immediate neighborhood of Schickard, toward the southeast. I refer to what, as far as its telescopic appearance goes, might be described as an enormous bubble—a bubble 54 miles in diameter. Unlike the other formations the surface of this singular ring is elevated above the general level of the moon. When we come to examine it in detail it hardly answers, perhaps, to my designation of a bubble, since the edges are a little higher than the center, giving it the form of a shallow dish. If we could visit it we should find on approaching that we were climbing the slopes of what would seem to be a chain of low mountains, and on reaching the summit we should see before us an elevated circular plain, sinking gradually toward the middle. Filled with water it would form a shallow lake lying on the top of a broad, flat mountain. There is nothing else quite like it on the moon and certainly nothing on the earth."

"It must have been a great curiosity in the days when the moon was inhabited, and I suppose that scientific 'lunarians' organized expeditions to explore it."

"Perhaps, if you choose to regard it in that way. Now look again at the *Mare Humorum*. You perceive that its eastern side is lined with mountains and crater rings, while near the center of the northern border there is a conspicuous ring with a bright line running from the southern edge to the center. This is one of the most beautiful of lunar formations, and is named Gassendi. It is a favorite object for those who study the moon with telescopes on account of the great variety and singularity of the details visible within the ring. When you become a selenographer and possess your own telescope you will find few things more interesting to study than Gassendi.

"Next let us take up photograph No. 12. Here the moon is once more a little 'older' than before, and the sunrise line has again advanced a little eastward. This advance does not appear so rapid when the terminator is near the moon's limb, because, on account of the rounding away of the lunar globe, the illuminated surface is foreshortened from our point of view on the earth. In this photograph you perceive that the wonderful shining mountain Aristarchus has become even brighter than it was before, or at least it is more conspicuous on account of the appearance of what seems to be a short ray shooting out from it in a southeasterly direction. There is also a light spot just below it which is caused by a little mountain group called the Harbinger Mountains. The bright ray connects Aristarchus with its neighbor Herodotus, of which I spoke a little while ago. There is a very remarkable feature of the moon here, not shown in the photograph, but to which I must briefly refer. It is an enormous cleft, or crack, or, if you please, cañon, which starts from Herodotus, whose northern wall seems to have been broken through to give passage to it, and goes winding across the surface of the *Oceanus Procellarum* with several sharp turns and angles for a total distance of nearly a hundred miles. What produced this remarkable chasm on the moon it is difficult to say. Some have suggested that it may once have been the bed of a river, but there are many serious objections to that view. Nevertheless, there seems to be little doubt that if we were to visit the moon we should find, in many ways, a striking resemblance between this prodigious cañon and that of the Colorado River."

"And are not all these things so ancient, as far as you can tell, that, like the terrible volcanic rings, they might have been formed before the appearance of inhabitants upon the moon?"

No. 12. September 4, 1903; Moon's Age 13.27 Days.

"They certainly seem to be very ancient, and I cannot deny the *possibility* of what you say."

"Very well, then, I, for my part, am convinced that curious eyes, filled with the light of intelligence, have peered down from the verge of that chasm into its fearful depths. If you will not permit me this flight of imagination I shall refuse to take any further interest in the moon."

"Oh, I should not think of refusing. Imagine what you will, and draw your own inferences, only remembering that they are not supported by *ascertained* facts, and probably never will be. Yet for all that they may have an element of truth."

"Pardon me for saying that your astronomical science, as far as it concerns the moon at least, does not seem to me quite satisfying. You are not bold enough in drawing conclusions."

"On the contrary many astronomers think that some of their brethren are altogether too bold in that respect. However, it must be freely confessed that astronomical science, except perhaps in its mathematics, is not satisfying even to those who have created it. Nobody would rejoice more sincerely than the astronomer at the discovery of evidence of the former, or even the present, habitability of the moon. It is surely a great disappointment that we have not been able to settle so apparently simple a question in regard to our nearest neighbor in the sky."

"Then if I were a multimillionaire I should certainly devote several of my millions to the construction of a telescope great and powerful enough to reveal so interesting a secret."

"With your great telescope you could probably render possible many discoveries at present beyond our reach. But the mightiest telescope that you could make would enable no one to *see* inhabitants on the moon, even if they existed."

"Not if it magnified the moon a million times?"

"No, for optical imperfections and the disturbances to clear vision produced by our atmosphere would absolutely prohibit the use of any such magnification. And even supposing that one could use a magnifying power of 1,000,000 diameters in viewing the moon, how near do you think that would place us to the lunar surface? It would still appear to be more than a quarter of a mile away."

"That is not much. I am sure I can see people at that distance."

"Oh, yes, but the distinctness of view would be nothing like so great as if you were looking at the same objects on the earth. Still, if we could obviate the atmospheric and other difficulties, a magnifying power of one million would certainly enable us to discover the works of the moon's inhabitants—their houses, their fields, their plantations, their great establishments of art and industry. But I assure you that a telescope of such power is a mere dream. It could never be constructed without some fundamentally new and unheard-of discovery in optics. We shall do better to turn once more to our photographs which, at least, have no deceptions. Dropping No. 12, we shall take up No. 13, which brings us practically to the Full Moon phase. The moon's age at the time this photograph was made was nearly fourteen and one-half days. You see that its whole eastward face is now lying in the sunlight. The march of day across its surface has been completed, and on the western edge of the moon the sun is about to set, while on the eastern edge it is just rising. Among the new things that have come into view is a conspicuous dark oval, shaped like Plato, but very much larger, near the eastern edge. This is a walled plain named Grimaldi, and it enjoys the distinction of being the darkest on the moon. Near it on the northeast and consequently closer to the limb is another walled plain, which I promised some time ago to point out to you because it bears the name of the astronomer Riccioli, the great bestower of names on the moon, and upon whose lack of imagination you have so severely commented. But, as you have already learned, the time of Full Moon is not the best for studying the mountains and rings, because then the light strikes too nearly vertical upon them and they cast no shadows. But it is the best time for seeing the broad general features of the lunar surface. Turn the picture upside down again, thus bringing the disk into its natural position as seen with the naked eye, and this photograph shows the moon very much as it appears with a small pocket telescope, or with a powerful binocular. The new prism binoculars that have come into use within the past few years are excellent for general views of the moon. Their defining powers are superb, and one who has never seen the moon with such a glass is always greatly surprised and delighted with the view which it affords. You see now that Tycho forms a blazing brooch, resting on the Maiden's neck, while its rays extend across her profile, and the long one lying over the *Mare Serenitatis* bears some resemblance to a pin displayed in her hair, with the crater ring, Menelaus, glittering at its lower end. The other bright point, to the left of Menelaus (we will henceforth keep the picture reversed), is a ring mountain named Manilius. After the detailed study which we have given to the various 'seas' and formations you should be able to recognize them with the picture in this position, and I wish that you should do so because, as I have just remarked, this is the position of the Full Moon as it is always seen with the naked eye or with a simple binocular, for the latter does not reverse it, as does a telescope. The western edge is now at the right hand, and the north at the top. All the *mares* are clearly visible. On the right the *Mare Crisium*, the *MareFœcunditatis*, the *Mare Nectaris* and the *MareTranquillitatis*; in the center, above, the *MareSerenitatis*; on the left the *Mare Imbrium*, the *Mare Vaporum*, the *Mare Nubium*, the *MareHumorum*, and the *Oceanus Procellarum*. The two bright spots on the right, lower than the *MareFœcunditatis*, are Petavius and a neighboring ring. Vendelinus forms a less brilliant spot at the western edge of an extension of the *MareFœcunditatis*, and Langrenus is distinctly seen on the western shore of the main body of that *mare*. Proclus and the remarkable diamond of the 'Marsh of a Dream' are very plain just under the large oval of the *Mare Crisium*. The mountains and cliffs encircling the *Mare Imbrium* on the west, north, and east you will recognize at a glance. The dark Plato is conspicuous in the lighter mountainous area north of this 'sea,' and the semicircle of the 'Bay of Rainbows' is sharply defined. Farther north is the long, dark *Mare Frigoris*, whose eastern end merges into the broad *Oceanus Procellarum*. Aristarchus appears as a very bright point in this 'ocean,' and far to the right of Aristarchus, toward the center of the disk, Copernicus, with its splatter of irregular rays, is conspicuous. Following the eastern limb round toward the south we see again the dark oval of Grimaldi, beyond which the bright mountainous region broadens as we approach the South Pole.

No. 13. September 5, 1903; Moon's Age 14.40 Days.

"There is just one other thing on which I should like to dwell a little while we have the Full Moon before us. I have already referred to it once or twice—I mean the system of bright rays or bands radiating from Tycho. These rays, as I have told you, are among the greatest mysteries of the moon. Their appearance is so singular and, if I may so describe it, unnatural, that when the first photographs of the Full Moon were published, some persons actually thought that they were being imposed upon. They imagined that the photographer had indulged in a practical joke, by photographing a peeled orange and dubbing it 'the moon.' The mysterious rays do not start from the central mountain of Tycho, nor even from the ring itself, but from a considerable distance outside the ring. Nevertheless, Tycho is manifestly the center from which they arise. It looks as though some irresistible force had been focused at that point—a force that split the moon along a hundred radiating lines. This is, in substance, the theory of the English selenographer Nasmyth. He supposed that, the lunar globe being burst by internal stress, molten lava welled up and filled the cracks. After solidifying this lava possessed a lighter color and greater reflecting power than its surroundings and thus gave rise to the appearance of long bands."

"Really, your moon history seems to me to be made up of extremely tragical chapters. But I am content as long as you put all these terrific events sufficiently far in the past to leave time for the moon to have enjoyed a different kind of history since they occurred."

"But," I said, "even if I grant what you wish, you must admit that the greatest tragedy of all succeeded."

"What do you mean?"

"I mean simply that your imagined lunar age of gold, when the moon was full of animated existences and beautiful scenes, has also become a thing of the past; and what geological cataclysm can be compared in tragic intensity with the disappearance of a world of life?"

"But that disappearance was gradual, was it not?"

"Very likely it was, if it depended upon the slow withdrawal of the atmosphere and water."

"Good! Then again I am fairly well content, for all things must have an end. The most beautiful life finally merges into old age and death. I think I have read that some of your *savants* predict that the earth will not always be a living world. All that I ask is that you leave room somewhere in your lunar history for an age of life on the moon."

"Very well then. As I have told you several times, Science does not positively forbid you to picture such an age if you will. She only says that she cannot find the evidences of its existence. Still, as we are going to see later, there are those who think that they can perceive indications of some simple forms of life on the moon even now. I will grant you that in the past these may have been more numerous and more highly organized."

By this time the afternoon had waned and the trees were lengthening their shadows upon the lawns of the park.

"Perhaps," I said, "we had better postpone an examination of the remaining photographs of the series exhibiting the moon's various phases until after dinner. They will show very well in the light of the electric chandelier. I have but a few words to add concerning the rays of Tycho. The opinion of Nasmyth concerning their mode of origin has not been universally accepted. Prof. William H. Pickering, for instance, has suggested that the rays are formed by some whitish deposit from the emanations blown out of comparatively minute craters lying in rows. He supposes large quantities of gas and steam given forth from craters surrounding the rim of Tycho, and, in consequence of these gases and vapors being absorbed and condensed in more distant regions, a wind constantly blowing away from Tycho and distributing the white deposit in windrows. A similar explanation has been applied to the shorter and more irregular systems of rays surrounding Copernicus, and a few other ring mountains."

"I prefer the Nasmyth hypothesis," said my friend, as we rose and took the path to the house. "It is, to be sure, more gigantically tragic, but then it is simpler and more easily comprehended."

III FULL MOON TO OLD MOON

III FULL MOON TO OLD MOON

AFTER dinner, in the brilliantly lighted drawing-room, we once more spread out the photographs on a table.

"This time," I said, taking up No. 14, "we are going to watch the advance of night over the moon. Before, it was the march of sunrise that we followed. Both begin at the same place, the western edge or limb of the moon. Comparing this photograph, which was taken when the moon was about fifteen and two-third days old, with No. 13, taken when the moon's age was more than a day less, you perceive, at a glance, wherein the chief difference lies. In No. 13 sunrise is just reaching the eastern limb; in No. 14 sunset has begun at the western limb. Having watched day sweep across the lunar world, we shall now see night following on its track. West of the *Mare Crisium* and the *MareFœcunditatis*, which I expect you to recognize on sight by this time, darkness has already fallen, and the edge of the moon in that direction is invisible. The long, cold night of a fortnight's duration has begun its reign there. The setting sun illuminates the western wall of the ring mountain Langrenus, which you will remember was one of the first notable formations of the kind that we saw emerging in the lunar morning. But then it was its eastern wall that was most conspicuous in the increasing sunlight. For the selenographer the difference of aspect presented by the various objects of the lunar world when seen first under morning and then under evening illumination is extremely interesting and important. Many details not readily seen, or not visible at all, in the one case become conspicuous in the other. But it is only close along the line where night is advancing that notable changes are to be seen. Over the general surface of the moon there is not yet any perceptible change, because the sunshine still falls nearly vertical upon it. Tycho's rays are as conspicuous as ever. Aristarchus, away over on the eastern side, is, if possible, brighter than before, and the three small dark ovals, Endymion a little west of the north (or lower) point, Plato at the edge of the *MareImbrium*, and Grimaldi near the bright eastern limb, are all conspicuous."

No. 14. August 26, 1904; Moon's Age 15.65 Days.

"But look!" exclaimed my friend, putting her finger upon the photograph. "Here is something that you have not mentioned at all. I believe that I have made a discovery, although you probably will not accept it as a scientific one. I see here a dark woman in the moon."

"I confess," I replied, "that I am not acquainted with her, and do not even see her. Please point her out to me."

"She appears in profile, like the brilliant Moon Maiden, but is not so much of a beauty. In fact I begin to suspect that she is the 'Old Woman in the Moon,' that I have often heard of."

"Positively I do not see her."

"Then I will try to recall some of the names that you have been telling me in order to indicate where you should. She faces west and occupies most of the eastern half of the disk. Her head is under Tycho, toward the northeast, I suppose you would say. The bright double ray that you pointed out in one of the preceding pictures lies across the top of her head and over her ear. Her face seems to be formed by a part of the *MareNubium*—you observe how well I have learned your selenographical terms—and her hooked nose is composed of a kind of bay, projecting into the bright part below Tycho. Her front hair is banged, and the *Mare Humorum* constitutes her chignon. She has a short neck, and a humped back, consisting of the *Oceanus Procellarum*. Copernicus resembles a starry badge that she wears on her breast, and Aristarchus glitters on the inner side of the elbow of her long arm. The *Mare Imbrium* seems to be a sort of round, bulky object that she carries on her knee, and she appears to be gazing with intentness in the direction of the *Mare Tranquillitatis*."

"Ah, yes," I said, laughing, "I see her plainly enough now. I really cannot say that your discovery is likely to be recorded in astronomical annals, but nevertheless I congratulate you upon having made it, if only for the reason that henceforth you can never forget the names and locations of the lunar 'seas' and other objects that you have been compelled to remember in pointing out your 'dark woman.' In truth, her features are almost as well marked as those of the Moon Maiden, but you will hardly be able to find her again, except in a photograph, or with the aid of a telescope, because you must recollect that this picture shows the moon reversed top for bottom as compared with her appearance to the naked eye, or with an opera glass. But please look again at the objects along the western edge, for we are about to turn our attention to photograph No. 15 in which this will be no longer visible. You must say 'good-by,' or rather 'good night,' to the *Mare Crisium* and the *Mare Fœcunditatis*; for you will see them no more, until another lunar day has dawned."

We next picked up photograph No. 15.

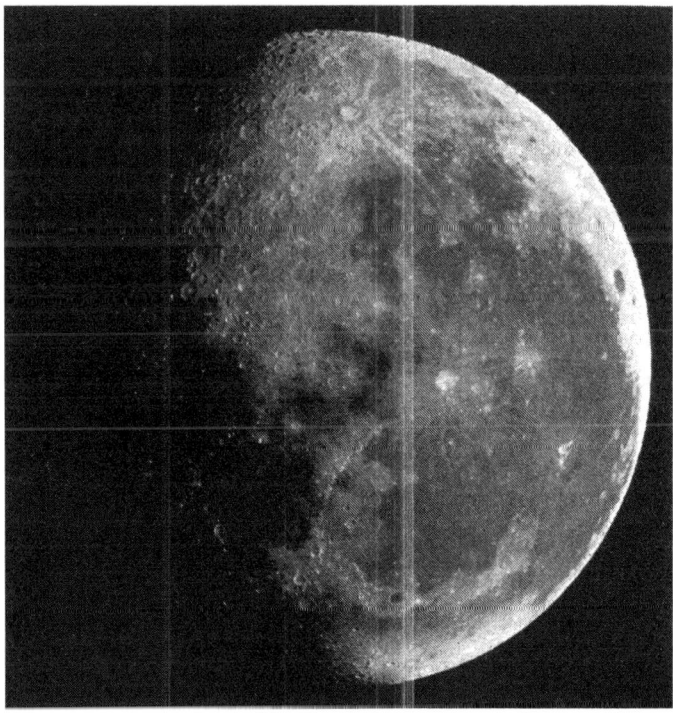

No. 15. August 28, 1904; Moon's Age 17.41 Days.

"Here the age of the moon has increased to nearly seventeen and a half days. The sunset line has advanced to the borders of the *Mare Nectaris* and the *Mare Tranquillitatis*. Toward the south a vast region which was very brilliant in the morning and midday light with the reflections from mountain slopes and the rays of Tycho, has passed under the curtain of night. The great crater rings on the eastern border of the *MareNectaris*, and thence upward to the South Pole, are beginning to reappear, but with the shadows of their walls thrown in a direction opposite to that which they assumed before. By a little close inspection you will recognize Theophilus and its neighbors which were so conspicuous for many days while the sunrise was advancing, but which have been almost concealed in the universal glare of the perpendicular sunshine since the Full Moon phase was approached. On the *Mare Tranquillitatis* and the *Mare Serenitatis* it is late afternoon, and your favorite 'Marsh of a Dream' has become a true dreamland."

"This oncoming of night," said my friend, "seems to me more imposing, and more suggestive of mystery than was the advance of day."

"Surely it is. Do we not experience similar sensations when night silently creeps over the earth? But it imparts a feeling of loneliness and desolation when we watch it swallowing up the barren mountains and plains of the lunar world that we do not experience in terrestrial life. There are no cheerful interiors on the moon to which one can retreat when darkness hides the landscapes. There is another thing about the lunar night to which I have made but scant reference thus far. I mean it's more than Arctic chill. Imagine yourself standing there in the midst of the broad plain of the *Mare Tranquillitatis*. Toward the east you would see the sun close to the horizon, yet blazing bright and hot, without clouds or mists to temper its rays. The rocks or soil beneath your feet would perhaps be cold to the touch, because the surface of the moon radiates away the heat very quickly, but your face and hands would be almost scorched by the intense solar beams. Looking toward the west you would see the shining tips of mountains suddenly extinguished, one after another, and when the sharply defined edge of the advancing night passed over you it would be as if you had plunged into a cold bath. In a little while, if you remained motionless, you would be frozen. No clothing would suffice to keep you warm. Nothing that polar explorers have ever experienced can be likened to the cold of the lunar night. Only the apparatus of the laboratories for producing temperatures, capable, when combined with pressure, of liquifying and solidifying the air itself, can bring about upon the earth a lowering of temperature comparable with that which occurs during the lunar night."

"But I do not exactly see why night should be so much colder on the moon than on the earth. She is not farther from the sun."

"No, her average distance from the sun is the same as that of the earth. The reason why her nights are so cold is to be found in the absence of an atmosphere like ours. The air is the earth's blanket, which serves a double purpose, tempering the heat by day with its vapors and winds, and keeping the earth warm at night by preventing the rapid radiation into space of the heat accumulated during the daylight hours. If there is any atmosphere at all upon the moon—and I shall tell you by and by what has been learned on that subject—it is so rare as compared with ours that it can exercise very little effect upon the temperature of the lunar surface.

"Now, look at the great range of the lunar Apennines. You will see that the eastern faces of these mountains are in the sunlight, and they cast no shadows, as they did in the lunar morning, over the *Mare Imbrium*. The same is true of the lunar Caucasus, and the lunar Alps. All of these mountains are very steep on the side facing the plains, and that is the side presented sunward in the lunar afternoon. By turning to photograph No. 16, we shall see this phenomenon more clearly displayed. This photograph, measured by the age of the moon when it was taken, is more than a day older than the other, but once again the effect of libration has, in part, counteracted for us the advance of the line of sunset. Still it has distinctly advanced. You will observe that it has now passed completely across the *Mare Nectaris*, and more than half across the *Mare Tranquillitatis*, while only the mountain tops along the western edge of the *Mare Serenitatis* remain to indicate its outlines in that direction. Theophilus, Cyrillus, and Catharina, on the eastern border of the *Mare Nectaris*, have again become very conspicuous, but this time in evening instead of morning light. See how sharply the western wall of Theophilus stands out against the darkness of night behind it, and how its central peak glows in the setting sun while all the vast hollow beneath it is black. The floors of Cyrillus and Catharina, being less profoundly sunken, are still illuminated. Below the *Mare Serenitatis*, the twin rings, Aristoteles and Eudoxus, are very conspicuous, and they show the same change of illumination as Theophilus, their western sides being strongly illuminated on their inner faces, while the eastern walls cast shadows into the interior. The mountainous character of the surface in the neighborhood of the North Pole of the moon seems to be more clearly brought out in evening than in morning light. In this picture the North Polar Region seems to be almost as much broken up with gigantic rings as is that surrounding the South Pole. In both cases, you observe, many of the rings are poised just on the edge of the lunar disk, and their libration alternately swings them in or out of view."

No. 16. August 29, 1904; Moon's Age 18.62 Days.

"Then the other side of the moon may not be very different from the side that is turned toward us."

"In its general features I doubt if it is at all different. There was once a theory, which had considerable vogue, that the side of the moon turned away from the earth presented a great contrast with its earthward side. A German mathematician, Hansen, drew conclusions, which are no longer accepted, as to the form of the moon. He thought that the moon was elongated in the direction of the earth, somewhat like an egg, her center of figure being about thirty miles nearer to us than her center of gravity. This, if true, would make the part of the lunar surface that we see lie at a great elevation as compared with the other part, and the center of gravity being toward the other side would cause the atmosphere and water to gravitate in that direction."

"What a pity that so interesting a theory should have been abandoned!"

"If interest were the only test of the value of a scientific theory knowledge would not advance very fast. Notice how this very photograph before us vindicates the true scientific attitude toward nature. It records all the facts within its range, and leaves the theories to us. The features of your 'dark woman' are, in their way, as clearly marked in the photograph as is the range of the lunar Apennines. It is for us to recognize the essential difference between the interpretations which we choose to put upon these two phenomena. Giving play to fancy, we see the figure of an old woman in the one case, and employing our reason we find a chain of unmistakable mountains in the other."

"But surely you do not mean to aver that science has no other business than that of recording facts."

"By no means. It is also the business of science to find hypotheses and to build up theories that will explain its facts and connect them together systematically, according to some underlying law. But as I have just intimated it is the mark of true science that it never retains a theory merely because it is interesting. The truth is the only touchstone. Still, even the most conscientious scientific investigator may be misled by his imagination. His greatest virtue is that he never lets his fancies deceive him after he has recognized their false character. Point out your 'dark woman' to the child, or the savage, and it will be in vain afterward to explain that her profile is made up of plains and mountains. The child and the savage are not scientific but imaginative, and only after a long education will they abandon the apparent for the real.

"I will ask you now to take up photograph No. 17. The age of the moon here is twenty days. Comparing it with the last photograph we see that Theophilus has disappeared, although Cyrillus and Catharina, being a little farther east, are yet visible. Half of the *Mare Serenitatis* is buried in night, and only a little of the eastern edge of the *Mare Tranquillitatis* remains visible. Aristoteles and Eudoxus are now very close to the terminator, and the shadows of their eastern walls are spreading farther over their floors. Aristarchus is very brilliant, as it is still early afternoon on that part of the moon, and the sunshine is intense. Observe that Kepler, the crater ring directly east of Copernicus, has become more conspicuous than we have seen it in any preceding photograph. This is especially true of the system of bright rays surrounding it, and it is due to the change of illumination. In the southern part of the moon, west of Tycho, you will now recognize many gigantic formations which we first saw when the sun was rising over them. Some of them are even more prominent in the sunset light. Among these is our old acquaintance Maurolycus, whose western wall is so brilliant that it resembles a tiny crescent moon. The double row of broad, dish-shaped walled plains along the central meridian has also become visible once more. In fact the amount of delicate detail and the sharpness of the definition in these photographs are very remarkable. Observe the curious mottling of the 'seas.' It is in some of the differences of tint, which correspond in telescopic views of the moon more or less closely with the varying shades in the photographs, that some selenographers have thought they could detect evidences of the presence of vegetation on the moon. We shall talk about that more in detail another time. It is sufficient just now to notice that the beds of the *mares* are by no means uniform either in tint or in level. All of them are more or less 'rolling,' like many of our prairies, and often winding chains of hills and huge cracklike ravines are visible in them. In this photograph the amount of detail shown in the *Mare Imbrium* is particularly striking. Notice how some of the crinkled rays from Copernicus extend almost to the center of the 'sea,' and how in front of the precipitous base of the Apennine range the lighter-colored ground, with three prominent ring plains in it, presents the appearance of shallows. Lying off the shore south of Plato and the Alps a number of isolated mountain peaks are seen, mere white specks on the gray background. The undulating character of the 'bottom' of the 'Bay of Rainbows' is also distinctly indicated. By the way, I should perhaps mention the names of the three rings lying off the front of the Apennines, for although they are among the most interesting on the moon they have hitherto escaped our special attention. The largest of the three is Archimedes, the second in size is Aristillus, and the smallest is Autolycus. You will hear of them again when we come to the large photograph of the *Mare Imbrium* and the *Mare Serenitatis*.

No. 17. October 10, 1903; Moon's Age 20.06 Days.

"Let me now prepare you for an almost dramatic change in the appearance of some of the most conspicuous lunar features which will take place when we pass from this photograph to No. 18. Direct your attention particularly to the chain of the Apennines. In No. 17 it lies very brilliant in the sunlight, with its western slopes distinctly visible, rising gradually from the shores of the *Mare Serenitatis* and the *Mare Vaporum*, while the 'sea' along its eastern front is bright with day. In No. 18 the Apennines have become simply a chain of illuminated mountain tips with comparative darkness all around them. Their western slopes are practically invisible, the *Mare Imbrium* on the east has turned dark, as if twilight had fallen over it—although as I have told you there is no twilight on the moon—and at its northern end the great range, with only its summits illuminated, projects like a row of electric lights far into the black night that has covered the plains beneath.

"Yet, although the *Mare Imbrium* has turned so dark as to be barely visible over its western half, the sun has by no means set upon it, and the darkness is perhaps greater than it should, theoretically, be under the circumstances. This phenomenon of the rapid darkening of the great lunar levels as the sun declines is one of the arguments that have been found to favor the hypothesis of the existence of vegetation. If, for the sake of discussion, we admit the possibility of vegetation growing on the lunar plains, it will be interesting once more to compare photographs Nos. 17 and 18."

"Don't say that it is merely for the sake of a discussion," interrupted my friend. "I shall be far more deeply interested if you will simply say that it may be true."

No. 18. September 29, 1904; Moon's Age 20.50 Days.

"Very well, let us put it that way, then. As I was remarking, if we again compare the two photographs, keeping the vegetation hypothesis in view, we may ascribe at least a part of the rapid darkening of the plain of the *Mare Imbrium* to a change in the color of the—what shall I say, grass?—covering it."

"Good! good!" exclaimed my friend, clapping her hands. "Just listen to him! After gravely rebuking me so many times for my unscientific faith in the lunar inhabitants of a long past age, now you are talking of 'grass' on the moon."

"You are hardly fair," I protested. "It is you who have just led me to make an admission which many astronomers would laugh at, and you ought to support me with all the brilliance of your imagination when I try to picture a state of things so consistent with your predilections about the moon."

"Oh, I do support you with all my heart!" she replied. "Pray go on, and tell me about the lunar grass."

"Not just at present," I said. "We are going to take that subject up again, and I may then succeed in convincing you that there is far more evidence for believing that vegetation exists on the moon in the present day than for believing that intellectual beings inhabited it at some unknown former period. I should warn you, too, that I have been using the contrasts of light and darkness between these two successive photographs simply as an illustration of what occurs in visual telescopic views; but that, for some reason, the lunar plains nearly always appear darker in photographs when contrasted with the mountainous regions than they do when viewed with the eye. Owing, also, to a variety of influences two successive photographs of the moon may differ in tone when the eye would detect no corresponding difference. All this, however, does not invalidate what I have said about the lunar 'seas,' or plains, darkening near sunset more rapidly than we should expect them to do, as a simple result of the low angle at which the sunlight strikes them.

"You will notice that the waning of day between photographs Nos. 17 and 18 has produced a remarkable change in the appearance of Tycho. Since the Full Moon phase Tycho has resembled a button rather than a volcanic crater, but now it has once more assumed the form of a very beautiful ring with its central peak clearly shown, its western wall, bright and its eastern wall casting a broad, black shadow. Most of the rays have now disappeared, only two or three, running over the eastern hemisphere, remaining visible. The immense walled plains near Tycho have again become prominent, Maginus toward the southwest, Clavius toward the south, and Longomontanus toward the southeast being the most conspicuous. Clavius is always a wonderful object for the telescope, but it is rather more interesting in the lunar morning than in the evening. Away over near the eastern limb, where the sun is still high, Grimaldi shows its dark oval, with a couple of mountain peaks on its western rampart shining brilliantly. The small, dark spot below it, toward the east, is in the walled plain, Riccioli. The bright spot with starlike rays, a long way south of Grimaldi, and east of the *MareHumorum*, is Byrgius, a walled plain near which exists a small system of bright streaks resembling those surrounding Copernicus and Kepler, but much less extensive."

"Do you recall my expression of impatience this morning when you were giving me the names of a long string of crater rings?" said my friend, smiling. "Well, I am now going to make a confession. Perhaps it is slightly of a penitential nature. I find now that these names, although they certainly are far from picturesque in most cases, begin to interest me, because, I suppose, I understand better the character and meaning of the things that they represent. The ceaseless Latin terminations no longer annoy me, for I do not think of them, but of the things themselves."

"It is always so," I replied, "whenever one takes up a new study. I know that you have dipped a little into botany, and I am sure that the Latin names which abound in that science must have repelled you at first. But after a time, when you had begun to recognize the beautiful flowers and the remarkable plants for which they stood, you found that even these names assumed a new character and became interesting and memorable. You will find it the same if you continue to study the moon. The most stupid designations will derive interest from their applications."

"Yes, that is no doubt true. Still, I wish that Riccioli had possessed a little more imagination."

"Be thankful, then, that he did not name the lunar 'seas' and 'bays.' You must now bid good night to your 'dark woman.' You observe that the *Mare Nubium* is beginning to fall under the shadow, and that her features are growing indistinct. If you will turn the photograph upside down you will find that the Moon Maiden has retired. She belongs exclusively to the western hemisphere, and it is only the eastern hemisphere of the moon that now remains visible to us, for we are close to the phase of Last Quarter. This is an aspect of the moon with which you may not be very familiar. To see the moon at Last Quarter, and particularly after she has passed that phase, we must rise near midnight and devote the early morning hours to observation. During these later phases, however, one may see the moon in the heavens during the daytime all through the forenoon and a part of the afternoon. She is a very beautiful object then, although few persons, I fear, ever take the trouble to look at her. The lighter parts of her surface assume a silvery tint in the daylight, and the dark plains seem suffused with a delicate blue from the surrounding sky. Exquisite views of the moon may then be obtained with a telescope. The glare of reflected light from the mountains and crater rings, which dazzles the eye at night, is so reduced that the telescopic image becomes beautiful, soft, and pleasing. The same principle has been very successfully applied in recent years to the study of the planet Venus. Her atmosphere is so abundant, in contrast to what we find on the moon, that she is as blinding in a telescope as a ball of snow glittering in full sunshine; but when seen in the daytime, her features, indistinct at the best, may be more clearly discerned."

"Oh, you interest me deeply! If Venus is supplied with such an abundance of air, I suppose she is inhabited?"

"It is not exactly orthodox among those calling themselves astronomers to talk of inhabitants on the planets, but I do not mind telling you privately that I think that Venus is most likely a world filled with all kinds of animate existences. Our present business, however, is with the moon, and I must recall your attention to the photographs. We shall next take up No. 19. Here the crescent shape becomes again evident, but reversed in position as compared with the crescent of the new and waxing moon. Only two of the 'seas' now remain completely in view—the *MareHumorum* and the *Oceanus Procellarum*."

"That term I think you have translated as the 'Ocean of Tempests.' Pray, do you know any reason why it should have been thus named?"

No. 19. August 16, 1903; Moon's Age 23.81 Days.

"There is not the slightest reason that I know of. You must ascribe it to the vivid imagination of that old astronomer whom you so greatly admire. I regret, sometimes, that he cannot be here to explain to you the thoughts that occupied his mind. They must surely have been very captivating, even though not very scientific. Remark that there are many of the features of the eastern part of the moon which we can now discern more clearly than in any of the preceding pictures. Beginning at the top we see the vast inclosure of Longomontanus with the top of its encircling walls illuminated, while the interior is all in deep shadow. Its western rampart projects into the night and seems detached from the main body of the moon. Along the terminator below Longomontanus, what appears to be another immense walled plain presents a similar aspect. This, however, consists of several smaller formations grouped near together, only their loftiest points being illuminated. The steep borders of the *Mare Humorum* are finely shown. Notice how the floor of that little 'sea,' which is about the size of England, as Mr. Elger has remarked, is mottled with whitish spots, and how distinct the ring of Gassendi appears at the northern end of the *mare*. You can even see the comparatively small crater that crowns the northern wall of the ring. Southeast of the *Mare Humorum* are visible the great flat plains of Schiller and Schickard. Notice also how all the surface of the moon in that direction is freckled with crater pits, which resemble the impressions made by raindrops in soft sand. But the smallest of these pits is larger than the greatest volcanic crater on the earth.

"The *Oceanus Procellarum* is beautifully illuminated in this picture. In several places, particularly north of the *Mare Humorum*, parts of *submerged* rings are visible. These are great curiosities, and we shall see more of them elsewhere. Some selenographers believe that they are the remains of an earlier world in the moon, which was buried by a tremendous upheaval and outrush of molten material from the interior. You will remember, perhaps, that I spoke of a catastrophe of that kind when pointing out the half-buried ring of Fracastorius at the southern end of the *Mare Nectaris*."

"Did that catastrophe occur after the formation of the huge lunar volcanoes?"

"It is difficult to say just when it occurred, but the appearances generally favor the view that it was subsequent to the great volcanic age. It is the opinion of Mr. Elger, whom I have once or twice mentioned as an English observer who has devoted special attention to the study of the moon's surface, that the *mares*, as we now see them, do not represent the original beds of the lunar oceans. These beds, which, according to this view, were at first deeper, have been covered up, at least over a great part of their areas, by the outrush of molten lava. If they were ever filled with water it was very likely prior to that occurrence. But you must remember that all this is speculation, very interesting, it is true, but based upon insufficient data to enable us to be sure of our conclusions. I shall show you later that some recent students of lunar phenomena have formed the opinion that there is a strong argument to be drawn from geological analogies in favor of the view that the lunar *mares*, practically in the state in which we see them, have been true sea beds.

"Let us continue our inspection of photograph No. 19, which is one of the most interesting of the series. Look at the crater ring Kepler, in the midst of the *Oceanus Procellarum*. We have not before seen it in the aspect which it now presents. Hitherto it has appeared only as a bright point surrounded by a light patch covered with radiating streaks. But now, with the late afternoon sunlight striking across it, its walls are illuminated in such a manner that its very perfect ring shows very clearly, about half of the interior lying in shadow, which serves to give it a striking relief. If we suppose a time when the *Oceanus Procellarum* was a real ocean, and when Kepler was an active volcano rising above its waters, its situation, far from all shores, would have been not unlike that of the great volcano of Kilauea in the Hawaiian Islands. In that case we might assume that the streaks around it represent ancient lava flows, which spread far about over the bed of the ocean. The same explanation would apply to the streaks and rays around Copernicus, and half a dozen other similar ring mountains.

"You will also observe that the afternoon slant of the solar rays has considerably changed the appearance of Aristarchus. Now for the first time the crateriform shape of that most remarkable mountain has become evident on account of the shadow in the interior. This shadow has almost reached the central peak which is the brightest part of the entire formation. You may be interested in the fact that the brilliance of the central peak of Aristarchus is so great that it stands in an order by itself, in what may be called the photometry of the moon's surface. Ten orders of relative brightness have been adopted to represent the various reflective powers of different parts and spots of the moon. I copy them from Mr. Elger's list. They are as follows:

"0° = Black (example, the shadows of mountains).

"1° = Gray black (example, darkest places in the walled plains of Grimaldi and Riccioli).

"2° = Dark gray (example, the floor of Endymion).

"3° = Medium gray (example, interior of Theophilus).

"4° = Yellowish gray (example, interior of Manilius).

"5° = Pure light gray (example, surface around Kepler).

"6° = Light whitish gray (example, walls of Macrobius).

"7° = Grayish white (example, Kepler).

"8° = Pure white (example, walls of Copernicus).

"9° = Glittering white (example, Proclus).

"10° = Dazzling white (sole example, the central peak of Aristarchus)."

"Really, I am greatly surprised by what you tell me," said my friend. "I would never have imagined that there were so many different neutral tints on the moon."

"You would be still more surprised," I replied, "if I could present to you a similar table of the different tints of color that have been discovered there. But I am not aware that any scale of lunar colors has been prepared. There are, however, various shades of brown, yellow, and green. Most of them are found in the *mares* and walled inclosures. Some of them appear to be variable, and some are only to be detected under particular illuminations."

"Are not such colors an indication of something living there?"

"It may be so—an indication, for instance, of the existence of 'lunar grass,' the mention of which so amused you a little while ago."

"Oh, it was not the 'grass' that amused me, but your unexpected way of introducing it. I *want* to be convinced that there is grass there, and a great many other things besides grass. But I am not yet satisfied concerning that unique peak in Aristarchus. 'Dazzling white' you say is its description in the scale of tints. That excites my curiosity immensely. I think you have told me already that it cannot be snow, but you have spoken of the possibility of crystals and of metal. Do you know, I like the idea of ascribing the phenomenon to metal. It recalls something that I read in childhood about the first discoverer of a silver mine in Mexico. As I remember the story, an Aztec hunter, chasing his game across a mountain, seized upon a bush to aid him, and the roots giving way disclosed a glittering mass of silver. Why not let me imagine that the peak of Aristarchus is composed of pure silver?"

"There is no harm in imagining that if you wish to do so. But then your imagination, or rather your knowledge, should go a little farther and recall the fact that silver does not remain dazzling bright when exposed."

"Ah, but you say there is no air, no water, no rains, no moisture on the moon. Under such circumstances might not a metal remain bright?"

"It is possible, but I hardly think that it would. It is likely that other corroding influences exist. A better explanation, I think, is afforded by supposing that the reflecting surface is simply composed of a rocky mineral, resembling in its power of reflection a mass of quartz crystals or imbedded planes of mica. There is no absolute impossibility involved in thinking that it may be simply white rock."

"Why not say marble—a gigantic Carrara mountain on the moon?"

"I fear that that would involve a geological history for the lunar world for which we have not sufficient warrant in observed facts. I prefer to assume a volcanic origin for the phenomenon. Since you are so interested in the mystery of Aristarchus I may add that a part of the floor and the inner side of the ring are also extremely bright, but not quite so bright as the central peak. That alone stands at the top of the scale. Putting the peak at 10°, Mr. Elger finds that the other brilliant parts of Aristarchus possess only 9½° of brightness. Yet the whole interior is so glistening that when the sunlight falls vertically it almost resembles the inside of a crystal cup, and details are hidden in the glare.

"Now please look at the 'Bay of Rainbows' in the photograph before us. Cape Laplace at its western end lies close to the terminator and appears as a minute speck of light. The great bow-shaped shore is clearly defined, the level surface within being very dark and the highlands around it comparatively bright. The crater mountain Bianchini you will recognize near the center of the bow. Several other similar crateriform mountains are visible toward the north and east. In this light the surface of the moon eastward from the North Pole appears as rough and broken with craters and crater plains as we saw in the earlier pictures that it is toward the west.

"Before directing our attention to photograph No. 20, let us return for a

moment to Aristarchus. When speaking of that formation a few minutes ago
I interrupted myself in order to give you the scale of tints on the moon, which
demonstrated the unique brilliance of the peak inclosed by the ring. I intended
to point out to you then the fact that in photograph No. 19 we see, for the first
time, not only the ring of Aristarchus but its curious neighbor Herodotus. A
light streak, which we observed in an earlier picture, seems to connect the two.
It is better, however, to notice this now because in turning from No. 19 to
No. 20 you will perceive once more a change in the appearance of Aristarchus
and its neighborhood. In No. 20 Aristarchus is distinctly more conspicuous.
The night has advanced during almost exactly twenty-four hours, having in
the meantime swept across the entire length of the 'Bay of Rainbows,' which
we now no longer see. If we had been using a telescope during that interval
we should have beheld a very interesting spectacle, for sunset on the 'Bay of
Rainbows' is quite as remarkable, although in a very different way, as sunset
on the Bay of Naples. The astronomer, seated amid the lonely gloom of his
observatory dome, and watching the change of light and illumination on the
surface of the moon, has many an hour of solitary enjoyment of aspects of
nature that are quite impossible on the earth, and that frequently lure him
into poetic meditations which find no place in his notebook."

"I am very glad to hear you say that. It enhances my opinion of the
astronomers, and convinces me that after all they are not so severely scientific
as they describe themselves."

No. 20. August 17, 1903; Moon's Age 24.84 Days.

"If they were," I replied, "or if all of them were, it would be a bad augury for the future of their science. Do not think that in occasionally seeking to restrain your imagination I wish to express condemnation of what, after all, is the noblest of human faculties. But again we are forgetting our principal business, which is with the facts. Aristarchus, as I have said, has undergone another distinct change of appearance from that which it showed before. The central peak is now covered by the shadow of the eastern wall, but still the reflection from the western wall alone is sufficient to make it the brightest spot on the moon. Herodotus, on the other hand, has become indistinct and the Harbinger Mountains are practically invisible, but we can detect the existence of the enormous chasm or cañon, which I told you once issues from the interior of Herodotus and goes winding nearly a hundred miles over the floor of the *Oceanus Procellarum*.

"Notice, also, how clearly visible three or four relatively small craters east of the 'Bay of Rainbows' have become, and how conspicuous are several large walled plains on the northern 'horn.' The dark level south of these formations and between them and the small craters has also a name which I have not before mentioned. It is the *Sinus Roris*, 'Gulf of Dew.' It connects the *MareFrigoris* with the *Oceanus Procellarum.* It is another legacy from your friend the imaginative astronomer."

"Then once more he receives my thanks for having done his best to make the moon an ideal world. It is always painful to have one's ideals destroyed."

"I hope that I have not been destroying any of yours."

"No, but at least you have caused a change in my impressions about the character of the moon. Henceforth there will be an element of terror as well as of unexpected grandeur mingled with my thoughts of the 'Queen of Night.'"

"That element will not be diminished by what I am about to point out. Look far over near the eastern border of the *Oceanus Procellarum*, directly east of Aristarchus. There you will distinguish the outlines of two or three vast submerged ring plains, which we may regard as relics of that earlier lunar world, which preceded the outgush of lava that Mr. Elger thinks covered the sea bottoms. Observe also the singular light streak that runs from Kepler, now barely visible at the edge of night, to a dark little crater, beyond which lies a bright point off the coast of the 'ocean.' South of this there are other submerged ring plains, one of which, named Letronne, has a high western wall, which forms in the picture a sort of promontory projecting from the southern border of the *Oceanus Procellarum*, almost directly north of Gassendi. The latter is very clearly shown at the lower end of the *MareHumorum*, the western side of which is in shadow, while its whole surface has turned very dark. On the southern horn of the crescent the ring plains, Schickard and Schiller, are still prominent, and the northern and eastern edges of the *Mare Humorum* appear more ragged with mountains and crater rings than before."

"And have all these mountains and craters names?"

"Not all of them, but many more, perhaps, than you suppose. On the whole visible surface of the moon about 500 objects, not including the 'seas,' have received names. It may surprise you to learn that the position of the most important of these objects has been ascertained with an accuracy which is still lacking in our determination of positions on the earth. In other words our charts of the moon are more exact than those of our own planet."

"That does indeed surprise me. I should have thought that, living on the earth, we could make very correct maps of it, while, as for the moon, two or three hundred thousand miles away, it seems to me not so easy to do that."

"It is mainly because we are on the earth that we find such great difficulty in making accurate maps of it. We cannot look at the earth as a whole, but we have to crawl over its surface, making measurements as we go, and afterwards translating those measurements into lines and angles on paper. Thus we are still uncertain about the precise distance between many important points on our globe, while for points on the moon no corresponding uncertainty exists. The moon hangs before us in the sky, with no clouds except those in our own atmosphere to obscure it, and it is only necessary carefully to observe the position of particular points, and with the proper instruments to measure their distance and directions from one another. But even this is not a thing that can be accomplished without much pains and much knowledge. The astronomer, no matter what field he chooses, is necessarily a hard worker, and his motto, above everything else, is accuracy. No one is more tempted than he by the sublimity and the extraordinary character of the objects of his study, to give rein to the imagination, and yet imagination is the thing of all others from whose vagaries he must most carefully guard himself. So you must not blame him too severely if he has not dotted the shores of the moon with cities, and populated its plains with industrious farmers."

"If you will permit me to wander a little aside from our photographic studies for a few minutes," said my friend, "I should like to ask you about two or three things concerning the moon which have long puzzled me. From my earliest days, living the greater part of the time in the country, I have heard that the moon exercises a decided influence over the weather, and over the growth of vegetation. I have neighbors who would never think of planting certain things except 'in the New of the moon'! Some will not cut timber except 'in the Old of the Moon,' as they say that the sap is drawn up by the moon's influence when she is growing. Is there really any truth in all this?"

"Not the least. At any rate there is no scientific evidence whatever for such statements, and no probability that they are based on facts. They are the result of faulty observation, misled by coincidences. It is *imaginable* that the light of the moon might have some influence upon vegetable growth if it were an original kind of light coming from the moon herself. But moonlight is only reflected sunlight, and when we examine it with the spectroscope we do not find that the rays of light in visiting the moon and returning thence to the earth have had either anything added to or anything taken away from them, except intensity. The total amount of light reflected from the moon upon the earth is estimated to be about $1/618000$ of the total amount that comes to us from the sun. Curiously enough the moon appears to reflect proportionally more heat than light, the amount of lunar heat received by the earth being reckoned at $1/185000$ of the amount coming from the sun. The popular idea that the moon affects the movement of sap in plants is equally illusory."

"But about the weather? I know people who believe that a change of the moon from one phase to another brings about a change of weather. Is that true?"

"Certainly it is not true. The moon is changing its apparent form all the time. There is no sudden alteration at any phase. The popular belief, however, has always been so firmly fixed that many investigations have been made to ascertain whether there is, in reality, any foundation for it. These investigations have shown that no measurable effect of the kind exists."

"And the Full Moon does not drive away clouds, as some assert?"

"Surely she does not. I will now tell you something that the persons who plant and sow and cut timber according to the phases of the moon, and who believe that she exercises a kind of magic control over the clouds, probably have never heard of, although if they knew it they might use it as an argument in favor of lunar influences. It is this: The alternate approach and retreat of the moon with respect to the earth, as she travels round her elliptical orbit, produce measurable, although slight, disturbances of the magnetism of our planet. The distance of the moon varies to the extent of about 30,000 miles. Now, if it could be shown that these magnetic disturbances were reflected in the character of the weather, then the supposed influence of the moon would be established. But that has not been shown, and if it were shown it would still be found that the phases of the moon had no relation to the fact, for the moon may be at its greatest or its least distance from the earth, or at any intermediate distance during any possible phase."

"You will, perhaps, think me very persistent in asking foolish questions, but there is one other on my mind that I should like to put, now that we have gone so far. It is this: I have read, since the great earthquakes at San Francisco and Valparaiso, and the great eruption of Vesuvius in the same year, 1906, that the moon has an influence over such things. Is this another unfounded popular superstition?"

"It is not a notion of *popular* origin at all," I replied. "It originated rather from scientific considerations, and there may possibly be a germ of truth in it, although it yet remains to be demonstrated, and the evidence concerning it is confusingly contradictory. You will recall, I trust, what has been said about the sun and the moon producing tides in the oceans. We have also seen that before our globe had assumed its present condition, while it was yet more or less plastic throughout its whole mass, and before the birth of the moon, great tides were produced in the body of the earth. The *tendency* to the production of such bodily tides still exists, and now that the moon has become a near-by attendant of the earth, she acts more effectively in this regard than does the sun. If the earth were still plastic the moon would produce bodily tides in it. In other words the earth would be deformed by the attraction of the moon. The question has arisen whether or not the tendency to the production of such tides, now that the earth has become rigid, may not disturb its crust sufficiently to induce earthquakes and volcanic eruptions. Some students of the subject have thought that they could detect evidence that this is the case. It has frequently happened that such phenomena have occurred on a large scale, at or very close to, the periods of New and of Full Moon. Those are the times, as we saw when we were talking of the oceanic tides, when the sun and the moon pull together. If all great eruptions and earthquakes occurred at these conjunctions there would be little doubt of the correctness of the theory. But, unfortunately for the clearness of our conceptions, this is by no means the case. There have been many earthquakes and volcanic outbursts when the sun and the moon were not thus combining their tidal attractions. Thus the evidence is found to be contradictory or inconsistent, and the question remains unsettled. It is, however, a very interesting one, and the time will come, it is to be hoped, when it will be answered decisively one way or the other."

After this digression we returned to the study of the photographs.

"Photograph No. 20, which we have just been examining," I said, "represents the moon at the age of about twenty-four days and twenty hours. The next, and the last of the series showing the moon in progressive phases, is No. 21. Here the age of the moon is about twenty-six days and twenty hours. It is the fast waning sickle of the Old Moon which we behold. You perceive that it is relatively uninteresting when compared with No. 20, because very little except the eastern limb is illuminated. Nearly all the great circular and oval formations and craters, and all the 'seas,' have passed into the lunar night. Only the eastern verge of the *Oceanus Procellarum* remains in sight, dulling the brilliance of the inner curve of the sickle. The dark walled plain above the center is Riccioli, and just below it appears Hevel, a smaller, but yet large formation, with a low central mountain. It is hardly worth our while to attempt to identify the other features shown in the photograph. They include none that we have previously studied. Yet this picture has an interest all its own because it is an excellent representation of the moon at a time when she is so near to the sun. Do not forget that, as I warned you when we began with the crescent of the New Moon, in these photographs the moon appears reversed top for bottom. Seen in the sky in the early morning this sickle would have its rounded edge toward the left hand and directed more or less downward, according to the position of the sun. A great deal of confusion exists in the minds of well-educated people concerning the position of the sickle of the New and the Old Moon. You have, of course, heard of the classic instances in which artists have drawn the New Moon with the concave side toward the sun! It is only necessary to remember that a line drawn straight from the center of the convex side of the sickle, whether it be the New Moon or the Old Moon, always extends directly toward the place occupied by the sun."

No. 21. August 19, 1903; Moon's Age 26.89 Days.

"There is," said my friend, "an interesting old superstition which I have
often heard—I suppose it must of course be a superstition—concerning 'wet
moons' and 'dry moons.' As I recall it they say that when the sickle of the
New Moon appears nearly upright in the sky that is a sign of dry weather,
because the moon is then like an overturned cup, but when the sickle has its
ends turned upward that is a sign of wet weather, because then the cup can
hold water. I suppose that these various positions of the moon actually occur,
but I do not know how they are brought about."

"The supposed influence of the position of the New Moon on the weather,"
I replied, "is too gross a superstition to be worthy of any notice, but the
different attitudes of the sickle are interesting. They arise from the changes
in the position of the moon as seen from the earth with respect to the direction
of the sun, and these changes depend in turn on the inclination of the moon's
path in the sky to the plane of the earth's equator as well as to the plane of
the ecliptic or the earth's orbit. The ecliptic has an inclination of about 23½°
to the plane of the equator, and the moon's orbit is inclined a little over 5° to
the ecliptic. The moon may, in consequence, appear more than 28° above or
below the equator. But since, as I told you in the beginning, the orbit of the
moon itself turns slowly about in space, the distance of the moon above or
below the equator is not constant. It may be only a little more than 18°. In

consequence of these changes of relative position the situation of the horns of the crescent moon varies. But you need never be in doubt as to what position they will occupy at any time if you will simply remember that a straight line drawn from the point of one horn to that of the other must always form a right angle with the direction of the sun.

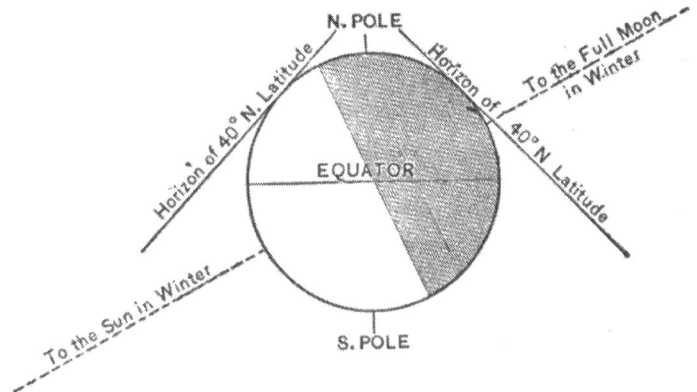

Diagram Showing Why the Winter Moon Runs High.

"There is another very interesting fact about the position of the moon in the sky which we should not neglect to notice. Did you ever observe the superior brilliancy of the light of the Full Moon in winter? It is one of the compensations that nature offers us. Since the Full Moon is necessarily situated opposite to the point occupied by the sun, and since the sun is far south of the equator in midwinter, it follows that at the same season the Full Moon appears high above the equator in the northern hemisphere. You will, perhaps, permit me to show you a diagram intended to explain this phenomenon.

You observe that the sun being south of the equator, in the direction indicated by the dotted line, the Full Moon is correspondingly situated north of the equator, and must necessarily appear high in the sky at midnight, when the sun is at its lowest declination. This is the reason why the winter Full Moons are so brilliant, making the snow-clad hills gleam with a splendor that sometimes dazzles the eyes of the beholders. In the Arctic regions the long winter night, when the sun does not rise for months, is periodically brightened by the presence of the Full Moon. Just the opposite condition of affairs exists in summer. Then the sun being north of the equator the Full Moon is south of it, and 'runs low,' appearing in high latitudes to skim along the southern horizon."

"Thank you, and now I will ask you one more question," said my friend. "I have often heard of the 'Harvest Moon' and the 'Hunter's Moon.' Will you not kindly explain what is meant by these terms and when the 'Harvest

Moon' can be seen? There is a poetic suggestiveness in the name that appeals to me."

"I will try with pleasure," I said, "but I fear that I shall have to trouble you with another diagram, or perhaps with two."

"Oh, I shall not mind that at all. I have grown used to diagrams as well as to the nomenclature of the moon."

"Well, if my diagrams conduct your thoughts to things as interesting as many that lie concealed behind the prosaic names on the moon I shall be content. To begin, then, I must remind you that in her monthly journey around the earth the moon moves from west toward east in her orbit, and thus she gets a little over 12° farther east every twenty-four hours, as reckoned from the position of the sun. The earth turning on its axis in the same direction causes the moon to appear to rise in the east and set in the west once every twenty-four hours. But in consequence of the constant eastward motion of the moon she rises at a later hour every night. Here is a graphic representation of what I mean:

"The earth is turning on its axis in the direction represented by the arrows, and simultaneously the moon is moving in its orbit in the same direction, as is shown by the large arrow. Suppose that some night the moon is seen rising at a particular hour from the point A on the earth. Then, the following night, when the observer has again arrived at A, with the rotation of the earth, the moon will have advanced from M1 to M2, and will not be seen rising until the point occupied by the observer has arrived at B. This retardation of the hour of moonrise is variable on account of changes in the position of the moon, arising from the inclination of her orbit to the plane of the equator, and from the inequalities of her motion, to which I have before referred. On the average it amounts to fifty-one minutes daily. It varies also with the distance of the observer from the equator, the variation being greater in high latitudes. In the latitude of New York the retardation of moonrise may be as great as an hour and a quarter, or as little as twenty-three minutes.

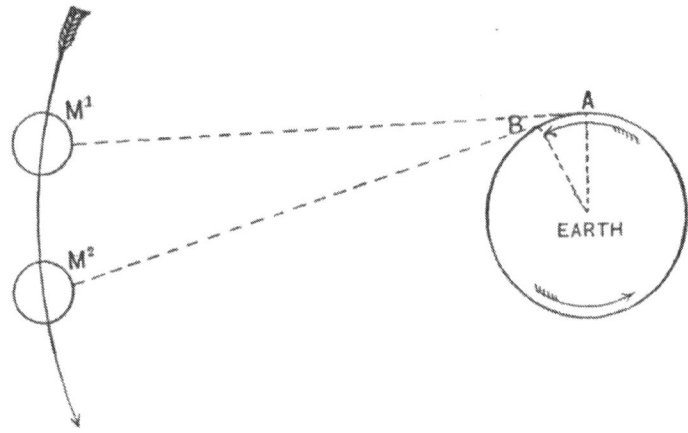

Diagram Showing Why Moon Rises Later Every Night.

"Now it is upon this variation that the phenomenon of the 'Harvest' and the 'Hunter's Moon' depends. If I had a celestial globe here I could show you that at the time of the Autumn Equinox, September 22d, when the sun crosses the equator moving southward, the apparent path of the moon in the sky intersects the eastern horizon at a comparatively small inclination. In other words the moon at that time instead of rising steeply from the horizon rises on a long slope almost parallel with the horizon. The consequence is that for several evenings in succession the Full Moon near the time of the Autumn Equinox may be seen rising just after sunset at almost the same hour. Look at this second diagram and you will see why this is so.

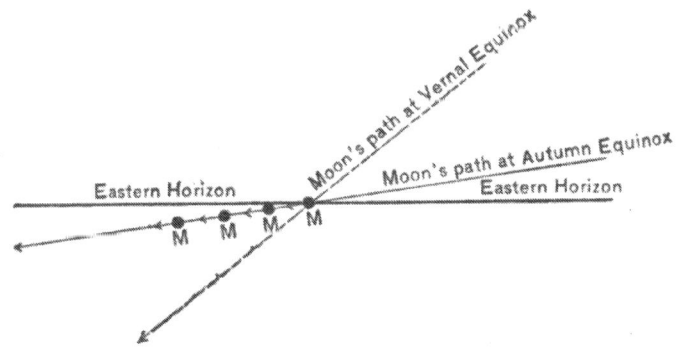

Diagram Illustrating the Harvest Moon.

"The little circles M show the moon at several successive positions in her orbit, just twenty-four hours apart. You perceive that in consequence of the slight inclination to the eastern horizon the sinking of the latter caused by the earth's rotation will bring the moon into view night after night at almost the same hour. In fact, in high northern latitudes like those of Norway and Sweden the moon's path at this time of the year may actually coincide with the horizon, so that for several evenings she will rise at exactly the same hour. The name 'Harvest Moon' explains itself, since it always occurs at the time of the autumn harvests and the vintage, and seems to supplement the fading daylight for the benefit of late laborers in the fields. The 'Harvest Moon' does not occur every year at precisely the same date. It is very rare that Full Moon happens to fall just on September 22d. It usually either precedes or follows that date. The 'Harvest Moon' is the Full Moon which occurs nearest to the Autumn Equinox, either before or after. The 'Hunter's Moon' is the first Full Moon which follows the 'Harvest Moon.' Like the former it rises for several successive evenings near the same hour, but this phenomenon is less marked in the case of the 'Hunter's Moon,' because it is farther from the Equinox."

"Thank you, again," said my friend. "I shall never henceforth look at the moon without thinking of circles, straight lines, and arrows as well as of 'ring mountains' and 'seas.'"

"Then you are making good progress toward science," I replied. "One last look, now, at the photograph of the Old Moon's sickle, and then we had better postpone our examination of the large photographs, showing certain particularly interesting districts on the moon, until to-morrow morning. There is here another interesting point for artists to note. The convex side of the sickle of the Old Moon, or the New Moon, is always an arc of a circle, but the concave side is never circular although it is often thus represented. The concave side, neglecting its irregularities arising from the differences of level and of brilliancy of the lunar surface, is elliptical in outline, that is to say, it is a semicircle viewed obliquely."

"Whatever its geometry may be," replied my friend, "it is certainly very beautiful. Good night, and I shall demand to see those large photographs before the sun is very high to-morrow."

IV GREAT SCENES ON THE MOON

IV GREAT SCENES ON THE MOON

MY friend did not leave me in doubt on the following morning as to the genuineness of her interest in her new studies. The shadows of the trees in the park were yet as long drawn out as the silhouettes of lunar peaks at sunrise, when we resumed our place under the elm, and, at her request, I opened once more my portfolio.

"The series of photographs that we are now about to examine," I began, "are on so large a scale that only a selected part of the moon is seen in each of them. But within the restricted limits of these pictures the amount of detail shown is truly astonishing, far more indeed than can be found on the most elaborate lunar charts. These photographs were made by Mr. Ritchey with the great 40-inch telescope of the Yerkes Observatory. Many more besides those that we are going to look at were taken by him, but I have selected, where choice was difficult, six which seemed to me to be of special interest. We shall begin with one which covers the larger part of the *Mare Nubium*, in the southeastern quarter of the moon. You certainly must remember the *MareNubium*, for it forms the head of the 'dark woman' whom you discovered in the moon last evening, and if you will hold this photograph at arm's length you will see that her face is unmistakably stamped upon it."

"I am greatly flattered," she replied, "that you should remember my discovery so well. I begin to feel hopeful that it may yet find a place in the books."

"It certainly is as deserving of such a place as many things that get into books. You ought to find a suitable name for this woman in the moon."

"If I believed myself capable of rivaling the man who christened the 'Marsh of a Dream,' I should surely try my hand at lunar nomenclature, but I fear that I should fall too far short of the ideal he has set up, and so I shall leave her nameless."

"Permit me then to continue to call her the 'dark woman' whenever a reference to her may seem useful in fixing the localities that we shall talk about in this photograph. The most striking object shown in the picture is the great ring mountain Bullialdus which forms an extraordinary ornament on the top of the 'dark woman's' ear. This photograph was taken when the line of sunrise ran just along the border between the *Mare Nubium* and the *Oceanus Procellarum*. The *Mare Humorum* is yet buried in night beyond the upper right-hand edge of the picture, but some of its bordering mountains and craters have been touched by the morning sunbeams. You will observe that a little more than half of the interior of Bullialdus—which, by the way, I did not mention by name when we were studying the series of phase photographs— is yet filled with shadow, but its double-headed central peak rises clear and bright in the sunlight. The shadow of this central mountain can be seen projecting toward the east over the floor. The east wall, which is distinctly terraced, lies in full sunshine, and the light streaming over the lofty crest of the western wall touches the floor on its eastern half. The steep outer slopes that lead up to the western rampart, and the deep parallel ravines cut near the crest are clearly shown. The distance across the ring from the summit of the wall on one side to that on the other is 38 miles. The depth of the depression is 8,000 feet below the crest of the walls, but the latter rise only 4,000 feet above the level of the *Mare Nubium* outside, so that Bullialdus is an excellent example of the characteristic form of the lunar volcano, which I tried to illustrate for you last evening. The central mountain is 3,000 feet high. East of the south point of the ring a shadow shows the existence of a profound cleft in the wall, while a little west of south appears a smaller crater ring very black with shadow, except on its eastern side. If we stood on the *Mare Nubium* and looked toward Bullialdus and its neighbor from a distance of 25 or 30 miles they would resemble a double, flat-topped mountain, with its serrated crests connected by a high neck. The summit of one of the little peaks shown in the photograph in the plain just west of Bullialdus would form an excellent point of observation. Still farther south stands another crater ring most of whose interior is also, at present, filled with shadow. East of this, and a little farther south, is still a third ring of similar aspect, from which a curious range of hills runs southward. Returning to Bullialdus you will notice the radiating lines of hills that surround it, and particularly a more lofty and broken range which runs eastward."

Bullialdus and the *Mare Nubium*.

"Bullialdus verily frightens me!" exclaimed my friend. "What an unearthly look it has! The longer I regard it the stronger becomes the indescribable impression that it produces. I begin to understand now what you meant when you promised to find a history in the moon. Truly there never can have been such another history. I almost feel that I do not care whether the moon ever had inhabitants or not. Its own story is more fascinating than that of any puny race of beings, passing their ephemeral lives upon its wonderful surface, could possibly be."

"I am glad," I replied, "that you have begun to enter into the spirit of those who long and carefully study the earth's satellite. You see now, that it is not necessary to the astronomer to find evidences either of former or of present life upon the moon in order to stimulate his zeal. For him, as you have yourself intimated, the relics of its past history, which this little world in the sky exhibits so abundantly, are of higher interest than any story of human empire, for they have an incomparably vaster theme. But to lighten our labor a little, let me once more refer to the 'dark woman,' whose features, like the outlines of a constellation, serve for points of reference. I began by remarking that Bullialdus seems to be placed just over her ear. Observe now that, taken together with its immediate surroundings, the great crater ring forms a kind of barbaric ear-ornament of most extraordinary form and richness of detail. The line of hills east of Bullialdus, of which I spoke a few minutes ago, connects the ring with a tumbled mass of mountains on the border of the *Mare Humorum*. These mountains run northward, or downward in the picture, for a distance of perhaps 150 miles, and then turn abruptly westward for a like distance; after which, in the form of a broken chain, constituting the eastern walls of a row of half-submerged ring plains, they change direction once more and run southward in the *Mare Nubium*. The whole system bears some resemblance to a gigantic buckle."

"What is that curious object below Bullialdus which resembles an old-fashioned gold earring?"

"I was about to speak of that. It is a ring plain named Lubiniesky, about 23 miles in diameter with a wall a thousand feet in height, except in the direction of Bullialdus where it is broken down. The interior is very flat, and it forms a fine example of the half-submerged lunar volcanoes which abound in this hemisphere. It may have had a central mountain like Bullialdus, but if so it has been completely buried under the influx of molten lava or whatever it was that covered this part of the moon. The perfect form of Bullialdus in all its details when compared with the mere outline that remains of Lubiniesky indicates that the former probably burst forth after the inundation of liquid rock that drowned the latter. Thus we have in these two neighboring formations two chapters of lunar history which, like the monuments of Egypt, tell the story of widely separated epochs. The row of still more completely submerged crater rings westward from Lubiniesky and Bullialdus show by their condition that the depth of the lava flood was probably greater in their vicinity than it was farther eastward.

"Now look southward from Bullialdus, at a distance about twice as great as that of Lubiniesky and you will see another partially submerged ring, with a more serrated crest. The name of this is Kies. It is remarkable for the lofty mountain spur which sets off from its southern wall, and also for the fact that one of the bright streaks from Tycho—one of a parallel pair that I pointed out to you last evening—traverses its flat floor and continues on, broadening as it goes, to a deep crater ring which we have already noticed, southeast of Bullialdus.

"South of Kies, at the edge of the *Mare Nubium*, is a lofty mountain range whose summits and slopes are very bright in the sunrise. At one point a great pass breaks through these mountains, leading to a sort of bay shut in on all sides by precipices and the walls of gigantic crater rings. The large crater ring at the eastern corner of this bay is Capuanus. The smaller ring on its western side with a conspicuous crater on its eastern wall is Cichus. Notice the fine shadow that Cichus casts, whose pointed edge is evidently due to the little crater on the wall. That 'little' crater is six miles across! The twin rings apparently terminating the mountain mass northeast of the bay are Mercator and Capuanus. Between these and Kies you perceive two short ranges of small mountains and then a kind of round swelling of the surface of the plain resembling a great mound. These formations are rare on the moon. They look like bubbles raised by imprisoned gases. The United States Geological Survey has discovered something similar in form, but infinitely inferior in magnitude, in the great mud bubbles that rise to the surface of the Gulf of Mexico off the mouth of the Mississippi River. But I do not mean to aver that the two phenomena are similar in origin.

"Near the southern shore of the *Mare Nubium* appears a long, dark line which starts at the edge of a crater ring, crosses the southern arm of the 'sea,' evidently penetrates the bordering mountains, and reappears traversing the dark bay near its northern edge, cleaving both walls of a small crater ring in its way.

"I should weary you, perhaps, with too much detail if I undertook to identify all of the prominent objects in this photograph. Returning to the southern shore of the *Mare Nubium*, I shall simply call your attention to the very large ring plain with terraced walls and a peak a little east of its center. This is Pitatus. An enormous ravine breaks through its eastern side and connects it with a smaller ring from which the dark line already mentioned starts. This dark line represents one of the most remarkable clefts on the moon. It looks as though the crust had been split asunder there over a distance of at least 150 miles. It bears some resemblance to the great cañon near Aristarchus and Herodotus, except that the latter is very tortuous and this is nearly straight."

"Have I not heard of something similar in connection with the California earthquake in 1906?" asked my friend.

"No doubt you are thinking of the great 'fault' which geologists have discovered off the Pacific coast of North America. There is perhaps some resemblance between these phenomena. Pitatus, I may add, is 58 miles in diameter. You will observe how its southern wall has apparently been broken down by the deluge of lava which buried so many smaller rings in the *Mare Nubium.* If you will now turn your attention to the left-hand side of the photograph, somewhat above the center, you will perceive a very strange object, the so-called 'Straight Wall.' It lies just west of a large conical crater pit which has a much smaller pit near its western edge. You might easily mistake the 'Straight Wall' for an accidental mark in the photograph. It is not absolutely straight, and near its southern end it makes a slight turn eastward and terminates in a curious, branched mountain, whose most conspicuous part is crescent-shaped. The wall is about 65 miles in length and 500 feet in height. It is as perpendicular on its east face as the Palisades on the Hudson. It is not a ridge of hills at all, but a place where the level of the ground suddenly falls away. Approaching it from the west you would probably be unaware of its existence until you stood upon its verge. The dark line that we see in the photograph is the shadow cast by the wall upon the lower plain. In the lunar afternoon the appearance is changed, and the face of the cliff is seen bright with sunlight. This curious object has attracted the attention of students of the moon for generations, and many speculations were formerly indulged in concerning its possible artificial origin. It has sometimes been called the 'Lunar Railroad.' Manifestly, whatever else it may be, it is not artificial. The closest analogy perhaps is with what we were speaking of a little while ago, a geological fault, that is to say, a line in the crust of the planet where the rocky strata have been broken across and one side has dropped to a lower level.

"The crater pit in the *Mare Nubium*, east of the 'Straight Wall,' is named Birt, and its twin, 75 miles farther east, is Nicollet. Look now at the hooked nose of your 'dark woman.' The huge wart upon it is a crater plain named Lassell. Between the lower end of the 'Straight Wall' and Lassell, and over the bridge of the 'nose,' a wedge-shaped mountain runs out into the *mare*. This is called the Promontorium Ænarium, and must have formed a magnificent outlook if ever a real ocean flowed at the foot of its cliffs. The ring with a crater on its wall below Lassell is Davy. You will note some very somber regions scattered over this part of the *Mare Nubium*. One of them forms the 'dark woman's' eye, and just over it, like an eyebrow, is a curving range of hillocks, including some little craters. On the 'cheek'—I am still utilizing the 'dark woman' as a kind of signboard—at the base of the 'chin,' appears a partly double range of large ring plains. The greatest of these, at the bottom, is named Fra Mauro, and you will notice within it a curious speckling of small craters. Adjoining Fra Mauro on the south are two intersecting rings, Barry being the name of the western and Bonpland that of the eastern one. The partially submerged ring is nameless, as far as I know, while the upper or southern member of the group, with a broad valley shut in between broken mountain walls opening out of its northern side, is Guerike. There is only one other object, on the extreme lower right-hand corner of the picture, to which I will ask your attention. It is a singular range of mountains thrown into a great loop at its northern end, and known as the Riphæan Mountains."

"It seems to me," said my friend, putting her elbow on the table, and leaning her head a little wearily on her hand, "that there is a great sameness in these lunar scenes—always crater rings with or without central mountains, always peaks and ridges and chasms and black shadows. Truly variety is lacking."

"But what could you expect?" I replied. "Is it not enough to stimulate your curiosity that you are looking intimately into the details of a foreign world? When you go to Europe you see there mountains, plains, rivers, lakes, cities, people, absolutely identical in their main features with what you see in America. But you find them endlessly interesting because of their comparatively slight differences from similar things with which you are familiar, because of the great age of many of the objects to which your attention is directed, because of the long course of history which they represent, and principally, perhaps, because you are aware of the sensation of being far from home. It ought to be the same for you here on the moon. These things that we are looking upon belong to a globe suspended in space 239,000 miles from the earth. If the features of our globe are practically the same everywhere, differing only in the arrangement of their details, you should not be surprised at finding that nature does not vary from her rule of uniformity on the moon.

"In the next photograph of the series," I continued, "we have a marvelous specimen of the lunar landscapes. It is perhaps the most rugged region on the moon. It includes two objects of supreme interest, Tycho, the 'Metropolitan Crater,' and Clavius, the most remarkable of the ring plains. You will no doubt recognize Tycho at a glance. It is near the center of the picture. Like the last photograph this one represents an early morning scene. The western wall of Tycho throws a broad, irregular crescent of shadow into the cavernous interior, but all of the eastern, northern, and southern sides of the wall are illuminated on their inner faces. The central mountain group is emphasized by its black shadow. A little close inspection reveals the existence of the complicated system of terraces by which the walls drop from greater to lesser heights until the deep sunken floor is reached. The diameter of Tycho is 54 miles, and it is at least 17,000 feet deep, measured from the summits of the peaks that tower on both the eastern and the western sides of its wall. The vast system of bright streaks radiating from Tycho is not seen here, the time when the photograph was made being too near the sunrise on this part of the moon. The dish-shaped plains crowded around Tycho form a remarkable feature of this part of the lunar surface. It would be useless to mention them all by name, and I shall ask your attention only to some of the principal ones."

"Thank you for being so considerate," said my friend, smiling. "I am sure that I should forget the names as fast as you mentioned them."

Tycho, Clavius, and their Surroundings.

"Oh, I have no fault to find with your memory," I replied. "I doubt if many selenographers could recall them without referring to a chart. Let us begin with the greatest of all, Clavius, which, you see, is near the top of the picture. I think I told you before that Clavius is more than 140 miles across. The great plain within the walls sinks 12,000 feet below the crest of the irregular ring, but the plateau outside, on the west, is almost level with the top of the ring. It is difficult to imagine a more wonderful or imposing spectacle than that which Clavius would present to a person approaching it from the western side, and arriving at about the time when this photograph was made, on the top of the wall. Notice how in one place the summit of a ridge, standing off on the inner side of the western wall, has come into the sunlight, and think of the frightful chasm that must yawn between. Clavius is so enormous that the two crater rings, each with a central mountain standing on its wall, seem very small in comparison with the giant that carries them, and yet they are 25 miles in diameter! Stretched out into a straight line, the tremendous wall of Clavius would form a range of towering mountains, extending as far as from Buffalo to New York. Look at the curved row of craters, the smallest larger than any on the earth, which runs across the interior. In addition to these there are many smaller craters and mountains standing on the vast sunken plain, some of them looking like mere pinholes, and yet all of really great size."

"Truly," interrupted my listener, "the giantism—I think that is the word you employ—the giantism of the moon appalls me! How can I ever think, again, that the so-called great spectacles of nature on the earth are really great? You have destroyed my sense of proportion. Such immense things standing on a world so small as the moon—why it seems contrary to nature's laws."

"I have already told you that the very smallness of the moon may be the underlying cause of the greatness of her surface features. And I may now add that if your imagined inhabitants ever existed they, too, may have been affected with 'giantism.' A man could be 36 feet tall on the moon and well proportioned at that, without losing anything in the way of activity."

"Indeed! You almost make me hope that there never were such inhabitants, for what beauty could there be in a human being as tall as a tree?"

"Very little to our eyes, perhaps. You recall the impressions of Gulliver in the land of the Brobdingnags. However, they are not my inhabitants but yours, and if the law of gravitation says that they must have been twelve yards tall, then twelve yards tall they were. Take comfort, nevertheless, in the reflection that, after all, we cannot positively assert that gravitation alone governs the size of living beings on any particular world. We have microscopic creatures as well as whales and elephants on the earth, and human stature itself is very variable."

"Thank you, again. You have saved my lunarians. And now please tell me what is that frightful black chasm above Clavius?"

"It is a ring plain named Blancanus, 50 miles in diameter, and exceedingly deep. It is so black and terrible because complete night yet reigns within it, except on the face of its eastern wall. It is really a magnificent formation when well lighted, but like so many other great things it suffers through its nearness to the overmastering Clavius. When Goliath was in the field his fellow Philistines cut but a sorry figure. Look at the marvelous region just below Blancanus and imagine yourself entangled in that labyrinth! You would have but a small chance for escape, I fancy."

"I am sure I should never have the heart to even try to get out of it. One might as well give up at once."

"Yes, you are probably right. But I will direct you to something not quite so frightful, although still very formidable in appearance. Still farther below you observe a huge ring plain whose eastern wall is brightly illuminated, while nearly all the interior plain, although comparatively dark in tone, lies in the sunshine. It is Longomontanus. I pointed it out to you in one of the smaller photographs. Longomontanus is 90 miles across and 13,000 or 14,000 feet deep, measured from its loftiest bordering peaks. The very irregular formation below it is Wilhelm I. It is remarkable for the mountainous character of its interior."

"For what William was it named?"

"I do not know. We are now near the southern border of the region that we inspected in the preceding photograph. In the lower part of this picture you perceive some of the projecting bays of the *Mare Nubium*, and you can see again the remarkable cleft of which I spoke. The large ring near the bottom of the picture is Pitatus with its smaller neighbor Hesiodus. It is from the eastern side of the latter that the cleft apparently starts. Pitatus, you see, has a central peak, while Hesiodus, as if for the sake of contrast, possesses only a central crater pit. The ravine connecting the two is plainly visible. Toward the east you will recognize again Cichus, with its crater on the wall and its broad shadow with a sharp point, while still farther east, on the very edge of night, yawns Capuanus. The two walled plains above Pitatus are Gauricus on the left and Wurzelbauer on the right. The hexagonal shape of the former is very striking. This is a not uncommon phenomenon where the lunar volcanoes and rings are closely crowded, and it suggests the effect of mutual compression, like the cells of a honeycomb. Away over in the northwestern corner is a vast plain marked by a conspicuous crater ring which bears the startling name of Hell. It borrows its cognomen, however, from an astronomer, and not, as you might suppose, from Dante's 'Inferno.'

"Before quitting this photograph permit me to recall you to the neighborhood of Tycho and Clavius. To the left of a line joining them you will perceive a flat, oval plain with a much broken mountain ring. This is Maginus. Last evening while we were looking at one of the smaller photographs I pointed it out under a more favorable illumination, telling you at the same time that it possessed the peculiarity of almost completely disappearing at Full Moon. Already, although day has not advanced very far upon it, you observe that it has become relatively inconspicuous. This is a lesson in the curious effects of light and shadow in alternately revealing and concealing vast objects on the moon. You will notice that in many particulars Maginus resembles a reduced copy of Clavius. But the walls of Clavius are in a comparatively perfect condition while those of Maginus have apparently crumbled and fallen, destroyed by forces of whose nature we can only form guesses. Evidently the destruction has not been wrought, like that of some of the rings in the *Mare Nubium*, by an inundation of liquid rock from beneath the crust. It resembles the effects of the 'weathering' which gradually brings down the mountains of the earth, but if such agencies ever acted upon the moon, then it must have had an atmosphere and an abundance of water. In any event, here before us is another page of lunar chronology. Maginus is evidently far older than Clavius; Clavius is older than the craters standing on its own walls."

We now took up the third of the large photographs representing a part of the southwestern quarter of the moon, more extraordinary for its mountains, plateaus, and extinct volcanoes than the famous southwestern region of the United States.

"Here is something that you will surely recognize without any assistance,"

I said. "In the lower left-hand corner of the picture is the great three-link chain of crater rings, of which Theophilus is the principal and most perfect member."

"Oh, I recall them well," replied my friend. "And yet they do not appear to me exactly the same as when I saw them before."

The Great Southwest on the Moon.

"One reason for that is because this photograph represents them on a much larger scale, and with infinitely more detail. Another reason is that now we are looking at them in the lunar afternoon instead of the lunar morning. We are going to see them represented on a still larger scale, presently, but there are many things in this picture well worthy of study. Advancing from the west, the line of night has fallen over the extreme eastern border of the *Mare Nectaris*, and the shadows thrown by the setting sun point westward. Observe how beautifully the brightly illuminated terraces and mighty cliffs of the western wall of Theophilus contrast with the black shadow that projects over half of the interior from the sharp verge of the eastern wall. The complicated central mountain is particularly well shown. The loftiest peak of this mountain mass, which covers 300 square miles, is 6,000 feet in height. You will see its shadow reaching the foot of the western wall. Theophilus is 64 miles in diameter, ten miles more than Tycho, and it is deeper than Tycho, the floor sinking 18,000 feet below the top of the highest point on the western wall. If it were the focus of a similar ray system it would deserve to be called the 'Metropolitan Crater' rather than Tycho. Plainly, Theophilus was formed later than its neighbor Cyrillus, because the southwestern wall of the latter has been destroyed to make room for the perfect ring of Theophilus.

"The interior of Cyrillus, you will observe, is very different from that of Theophilus. The floor is more irregular and mountainous. The wall, also, is much more complex than that of Theophilus. The broken state of the wall in itself is an indication of the greater age of Cyrillus. On the south an enormous pass in the wall of Cyrillus leads out upon a mountain-edged plateau which continues to the wall of the third of the great rings, Catharina. This formation seems to be of about the same age as Cyrillus, possibly somewhat older. Its wall is more broken and worn down, and the northern third of the inclosure is occupied by the wreck of a large ring. Observe the curious row of relatively small craters, with low mountain ranges paralleling them, which begins at the southwestern corner of Cyrillus and runs, with interruptions, for 150 miles or more. South of this is a broad valley with small craters on its bottom, and then comes an elongated mountainous region with a conspicuous crater in its center, beyond which appears another valley, which passes round the east side of Catharina, where it is divided in the center by a short range of hills. The southeastern side of this valley is bounded by the grand cliffs of the Altai Mountains, which continue on until they encounter the eastern wall of the great ring of Piccolomini, whose interior appears entirely dark in the picture, only a few peaks on the wall indicating the outlines of the ring. The serrated shadow of these mountains, thrown westward by the setting sun, forms one of the most striking features of the photograph. The northeastern end of the chain also terminates at a smaller ring named Tacitus. You see that Riccioli was rather cosmopolitan in his tastes, since he has placed the name of a Roman historian also on the moon. Beginning at a point on the crest of the Altai range, south of Tacitus, is a very remarkable chain of small craters, which extends eastward to the southern side of a beautiful ring plain with a white spot in the center. This ring is named Abulfeda. The chain of small craters or pits to which I have referred continues, though much less conspicuous, across the valley that lies northwest of the Altais. It is a very curious phenomenon, and recalls the theory advocated by W. K. Gilbert, the American geologist, that the moon's craters were formed not by volcanic eruptions but by the impact of gigantic meteorites falling upon the moon, and originating, perhaps, in the destruction of a ring which formerly surrounded the earth, somewhat as the planet Saturn is surrounded by rings of meteoric bodies, which may eventually be precipitated upon its surface. The moon is more or less pitted with craterlets in all quarters, but there are places where they particularly abound. On inspecting this photograph carefully you will perceive several rows of much larger pits, two or three of them in the upper half of the picture, and one below the center, crossing the little chain of pits that I have just mentioned. The linear arrangement of some of the ring plains is also very striking. In regard to the theory that the lunar craters were formed by the impact of falling masses I may mention that two distinguished French students of the moon, Messrs. Loewy and Puiseux, have lately expressed the opinion that all of the features of the lunar surface are most readily explicable as the result of causes similar to those which have produced the topography of the earth. If that is so there is no need for us to invoke the agency of meteorites in pitting the surface of the moon. South of the Altai Mountains you will see a singular collocation of ring plains and craters which somewhat resemble in their arrangement Theophilus and its neighbors. First comes a large sunken plain just above the mountains. In fact the Altai range constitutes the northwestern wall of this formation, which you may recognize by a conspicuous oval crater near its upper side. Above this broken ring appear three other smaller ones, grouped at the corners of a triangle. The one on the right, with a central pit and a small ring plain on the inside of its western wall, is called Zagut. Its close neighbor on the west with most of its interior in shadow, is Lindenau, remarkable for its depth. The most southerly and largest of the group, with four or five large crater pits forming a curved row across its interior, is named Rabbi Levi. Starting from the east side of Rabbi Levi there is a long row of similar craters rather larger than those in its interior, which runs eastward almost to the edge of the photograph. North of these, parallel with and, in some instances, touching the crater pits, is an equally remarkable row of flat, smooth, walled valleys, which seem to overlap one another on their western sides, and which increase in size the farther east they go. The largest of these, with a very irregular wall, and having a smaller ring with a central peak apparently attached to its northern side, is Gemma Frisius."

My friend had listened to me in silence for a long time, following my finger as it pointed out the various objects on the photograph, but now she interrupted again: "You were pleased to compliment my memory a little while ago," she said, "but do you really think that I can ever recall all this that you have been saying, with theories about huge flying stones hitting the moon, and a string of the strangest names that I have ever heard applied to objects that are no less bizarre?"

"Pardon me," I replied, "but you will remember more than you think you will. The very oddity of these Hebraic and Arabic names will serve to fix them in your memory, so that you will at least recognize them when you see them again. Those curious objects will also come before your mind's eye whenever you think of, or look at, the moon. Trust me when I tell you that you are forming a better acquaintance with selenography than you are aware of. As to the theory that I have mentioned, what can appeal more powerfully to the imagination than the idea of the moon being bombarded by the fragments of an immense ring falling from the sky? The fact that men of science have believed such a thing possible ought to form a strong appeal to your lively fancy. In any case, I am disposed to be merciless just now, like a man who has found a patient listener to his hobby, and I am going to trouble you with a few more odd names and singular facts."

"Well," she replied, with a sigh, ending with a smile, "go on. After all I believe I am really interested."

"I am sure you are, for who could fail to be interested by things so remarkable in themselves, and so vastly beyond all human experience, as those that this photograph shows? We stopped at Gemma Frisius. Let us use that for a new starting point. A considerable distance south, say about a hundred miles, is an old friend of ours, Maurolycus. It is the large ring plain, with another half obliterated, on its southern side, in the upper part of the picture. Notice the row of wrecked rings, beginning at a great crater on its northeast wall and running westward. The broad, flat plain directly east of Maurolycus is Stöfler, whose name you will also recall. I shall not trouble you with the names of all the rings south of Stöfler and Maurolycus, but simply ask you to observe that they form a winding row which leads to a very grand ring almost entirely buried in night, the inside of its western wall alone being bright with sunshine. This wall, and some mountain peaks near it, resemble brilliant islands lying in the edge of the Cimmerian ocean whose ethereal waves wash the broken coast of the moon. Follow the ragged sunset line downward, and all along you will see these islands of light in the darkness; tips of mountains still shining while the sun has set upon all the valleys around, somewhat as you have seen the snowy top of Mont Blanc and the pinnacles of its attendant giants glowing after the shades of night have fallen deep upon Chamounix.

"Look next, if you please, at the right-hand side of the photograph. Somewhat above the center, three conspicuous dish-shaped ring plains are seen, two near together, the third farther away toward the left and downward. The largest of these is Aliacensis, its near neighbor is Werner, and the third is Apianus. They are from 40 to 50 miles in diameter. Still lower, and nearer the middle line of the picture, is a row of four or five ring plains, varying from 30 to 40 miles in diameter. The uppermost, or most southerly of these is double, or, in fact, partly triple, for the lower member of the pair has a broken plain attached to its southeastern side. This one, with a small central peak, is named Abenezra. Its close neighbor on the southwest is Azophi. You notice the singularity of the names. The next one below, with a small crater on its east side, is Geber. Then comes Almamon, and finally, largest of all, Abulfeda, which I pointed out to you as marking the end of the curious row of little crater pits, running eastward from the Altai Mountains. There is just one other formation to which I wish to call your attention in this remarkable photograph, and then we shall turn to the next in the series. West of Abenezra and Azophi, about half way to the Altai Mountains, you will notice a very irregular depression with three strongly marked craters within it. This bears the name of Sacrobosco, an old-time astronomer. Its eastern wall with its shadow looks like an elongated letter W standing on end. Sacrobosco and its surroundings constitute one of the most intricate regions on the moon, high plateaus alternating with great sunken valleys, rings, craters, and crater pits. The wall of Sacrobosco is extremely irregular in height, shooting up in some places with peaks of 12,000 feet elevation, and sinking in others almost to the level of the surrounding plateaus."

We now took up the next photograph representing Theophilus and its companions on a greatly enlarged scale. My friend uttered a cry of astonishment upon seeing it.

"Dear me," she exclaimed, "the moon becomes more terrible every moment! Positively, I almost shrink from the sight."

The Giant Ring Mountains, Theophilus and its Neighbors.

"Yes," I assented, "it surely is terrible here. In a little while, however, I shall show you a lunar scene of surpassing beauty. But study this spectacle with an inquiring mind and you will find that it, too, has its attractions. You are now looking upon Theophilus, Cyrillus, Catharina, and the surrounding region as the astronomer sees them with the most powerful telescopes. Indeed, with the telescope he sees the details more sharply than they are visible here, for the best photographs still lack something in distinctness. The illumination when this picture was taken was practically the same as in the last that we examined, but the magnification is much greater. Look, now, at the central mountain in Theophilus. Its great buttresses cast their shadows into profound ravines and chasms, imparting to it a most singular outline. Observe the tooth-shaped shadows of its two principal peaks, thrown westward across the floor, while the broad shadow of the western wall emphasizes the immense depth of the depression. The glare of the afternoon sun on the cliffs of the inner side of the eastern wall is so brilliant that the details are obscured. But the surface of the moon outside, particularly toward the north and the west, is beautifully brought out with all its wonderful modulations and irregularities. Judging by appearances, those who hold that Theophilus and similar formations, notwithstanding their enormous magnitude, are really of volcanic origin, have the strongest reasons for their opinions. Immense flows of lava seem to have taken place on all sides of the great ring, entering the *Mare Nectaris* on the west. Notice the huge mountain fold which runs from the parallel ridges on the southwestern side of Theophilus to the crater ring Beaumont, lying west of Catharina. Observe, also, the complicated form of the wall dividing Theophilus and Cyrillus. Two deep ravines, shown by the shadows that fill them, cross one another like the arms of a flat letter X. One of these ravines turns northward along the wall and re-enters Theophilus, while the other continues for a long distance within the western side of Cyrillus. I cannot imagine a more interesting or a more stupendous excursion for a geologist, a mountaineer, or a seeker after wonderful and sublime aspects of nature, than a climb around the crest of the wall of Theophilus—if indeed such a climb can be regarded as humanly possible.

"Now, again, I am reminded of what I once told you about the amazing contrasts of light and darkness, and of heat and cold, upon the moon. Suppose yourself standing on the verge of the eastern wall of Theophilus where the edge seems sharpest, and looking down into the abyss at your feet. The sun's rays would be unbearably hot where they touched your face and hands, but if you let yourself down a little way into the blackness beneath you would not only pass instantly into night, but you would shiver and shrink with cold so frightful that no winter experience that you have ever had could give an idea of its intensity. From that point of observation you would look across a chasm of inky darkness, 25 miles broad, and see, towering up from the illuminated plain afar off, with their summits more than two miles below your level, the brilliant group of the central peaks, while behind them the crest of the western wall would appear like a bright line on the horizon 60 miles away. Changing your place to one of the peaks on the dividing wall you would look down into Theophilus on one side and Cyrillus on the other. Then upon lifting your eyes to the black, airless sky you would see the stars sparkling on all hands, and, hanging in the heavens like a portentous, strangely colored moon many times larger than the disk of the sun, would appear the mottled orb of the earth. The terrific nature of the scenery around you, the meeting of day and night at your feet, and the incredible blending together of their characteristic aspects in the sky above you, the startling magnitude of the suspended earth—all these things combined would make you feel as if you were not only in another world but in another universe."

"I no longer wish to visit the moon," interrupted my friend, shaking her head.

"Not if you were assured of a safe return?"

"No, it would upset my mind. I am certain that I should go crazy in such a world where everything seems to be topsy-turvy."

"Wait until we arrive at the 'Sea of Serenity' once more, and perhaps you will think better of it. Notwithstanding the increased magnification, the details in Cyrillus and Catharina are hardly better seen in this photograph than in its predecessor, but the increase of size is very effective in emphasizing some of the features of the surrounding district. Cyrillus is seen to have a decided hexagonal outline, and west of its southern corner is an exceedingly curious formation, approaching closely to a square shape. The wall is illuminated within on all four sides, and out of the midst of the lozenge-shaped shadow resting over the bottom of the included valley, rises a mountain which, like the walls, is bright with sunshine. On the southwest a semicircular ridge runs out into the darkness, its top brightly illuminated. The general effect of the entire formation is fantastic. And could you imagine a wilder scene than that presented by the elongated mountain mass, which starts from the southwestern side of Cyrillus, skirts the border of Catharina, and continues on along the northwestern side of the broad valley in the upper part of the picture? See how it has, apparently, been rent apart by tremendous forces and torn by volcanic outbursts, which have left yawning craters everywhere. Even the valley itself seems to be simply a chain of wrecked crater rings of vast size, the cross walls having nearly disappeared. Observe, too, the immense number of crater pits of all sizes scattered everywhere, both inside the ring plains (Theophilus alone having few of them) and over the surrounding country. We shall see a still more remarkable example of this pitting of the lunar surface in the neighborhood of Copernicus, which is the chief object in the next photograph that we take up."

We came now to the large picture of Copernicus, and my friend took it in her hands to examine it.

"It is a marvelous thing to look upon," she said, "but it doesn't frighten me as Theophilus did."

"No, Copernicus is rather sublime than terrifying in aspect. Its comparatively lone situation, with the *Mare Nubium*, the *Oceanus Procellarum* and the *Mare Imbrium* surrounding it on all sides with their broad, level expanses, gives it an appearance of solitary grandeur belonging to no other single formation on the moon. 'The monarch of the lunar ring mountains,' Mr. Elger has termed it. First let me tell you the principal facts known about Copernicus. It is 56 miles in diameter, two miles more than Tycho, and eight less than Theophilus. It is not as deep as either of those formations, the highest points on its walls being 12,000 feet. But the walls are more uniform in height than is usual with so extensive a ring. They are very steep on the inside, especially near the top, where their slope has been estimated by Neison at from 50° to 60°. To a person standing on their verge they would seem almost perpendicular. The central mountain consists of five principal peaks. The outer slopes of the ring are also steep, but its maximum height above the surrounding surface does not exceed 3,000 or 4,000 feet, so that Copernicus, like the other great ring mountains, is, in reality, a vast sink, encircled with a mountain ridge. You will note that Copernicus clearly exhibits the tendency to a hexagonal form which we have observed elsewhere, although it stands alone with no other great rings pressing against its walls. Curiously enough the form of Copernicus is very closely repeated in the small crater ring Gay Lussac, situated in the mountains on the lower (north) side. This picture, I should remark, unlike the last two preceding it, was taken near lunar sunrise, and accordingly the light comes from the west. This is the best illumination for studying Copernicus and its vicinity. Of all the great ring plains Copernicus perhaps gives the most striking testimony in favor of the view of those who hold that the lunar volcanoes were once the actual centers of volcanic action, resembling the volcanoes of the earth in the ejection of vapors, ashes, stones, and streams of lava. The slopes around Copernicus for many miles look as though they had been covered with lava and pitted with minor craters such as appear on the shoulders and in the vicinity of many of our volcanoes, while the appearance of the great ring does not contradict the theory of Nasmyth and Carpenter, which I have previously mentioned, that it was built up by ejections from a central crater now more or less completely filled. As I have already told you the lunar volcanoes differ essentially from those of the earth in that their central depressions lie deep beneath the level of the surrounding surface of the moon. This is strikingly true of Copernicus, and it is a result that might have been foreseen from the enormous size of the craters. A mountain of sufficient magnitude to carry the vast cup of Copernicus on its head, as Vesuvius, Etna, Cotopaxi, and Popocatepetl carry their craters, could not stand even on the moon. Observe the generally radial arrangement of the lines about Copernicus, recalling the similar arrangement of lava flows about terrestrial volcanoes. Some of these lines, as you will see, consist of long rows of pits. Similar phenomena may be seen along the lava streams that we are familiar with on our planet, where small craters break forth one after another. A striking example of this arrangement is visible in the photograph on the northeastern slope leading up toward the Copernicus ring. But you will also see many very remarkable rows of pits in the vicinity of Copernicus which are not radial in arrangement with respect to the ring. The most conspicuous of these is on the northwestern side, about half way between Copernicus and the ring of Eratosthenes, which standing at the upper end of the chain of the Apennines appears at the left-hand edge of the picture. There are hundreds, probably thousands, of these pits on all sides of Copernicus.

"One of the explanations that has been suggested for them is that they were produced by the fall of enormous volcanic bombs thrown from Copernicus when it was in eruption. I wish merely to mention this idea without comment. It however calls up another interesting theory, which has not met with much acceptance, to the effect that such lunar volcanoes as Copernicus may have been powerful enough to eject masses of lava and rocks with a velocity sufficient to enable them to escape from the attraction of the moon, whereupon they became meteorites traveling in independent orbits around the sun. Some of these, the theory suggests, may be among those that have fallen upon the earth. A velocity of a mile and a half per second would be sufficient to overcome the gravitation of the moon. That is only three or four times the initial velocity which some modern guns are capable of imparting to their projectiles."

"I am sorry," explained my friend, "that you seem to attach little importance to so interesting a theory. It stirs my imagination to think of the moon sending bits of herself back to her mother planet. For my part, the theory does not seem to be any harder to believe than that of your Professor Darwin that the whole moon was thrown off from the earth. Besides, it intensifies my appreciation of the grandeur of Copernicus when I am told that that great volcano could once bombard the earth across—what is it, 240,000 miles?—of space."

"As you always choose the most picturesque theories to rest your belief upon, I shall not complain if you accept the lunar volcano theory of meteorites," I replied. "But, for the present, we have done with it, and I am now going to ask you to inspect the photograph for other interesting objects. East and north of the great ring you will see an extensive mass of mountains. Those on the north, with immense buttresses projecting into the *Mare Imbrium*, are the lunar Carpathians. I have already directed your attention to a comparatively small crater ring which resembles a reduced copy of Copernicus, situated in these mountains at the head of a bay which penetrates southward between high ridges, for about 30 miles. This crater is named Gay Lussac. It has a small deep neighbor on the southwest. West of Gay Lussac the Carpathians gradually dwindle away until they sink to the level of the plain. Toward the east they project in several bold headlands, terminating with towering peaks into the 'sea.' Lying off the point of the headland on the western side of the bay that leads to Gay Lussac you will perceive two charming little craters, almost perfect twins. Much farther toward the north and west is a larger crater, more than half of whose interior is black with shadow. This is Pytheas. Its lonely situation is very striking, but upon close inspection you will notice that a low range of hills appears to connect it with the twin craters that I have just pointed out. This range of hills, lying on the 'sea' bottom, is curiously forked, the other branch leading to a pair of small peaks in the 'sea,' which possess no craters. The little crater east of Pytheas is also a beautiful object in the picture.

"Near the eastern end of the Carpathians the mountains make their greatest advance into the *Mare Imbrium*, leaving a large square-cornered bay on the west. From this point they turn southward, forming a complicated mass of peaks and ridges interspersed with craters and pits. These mountains east of Copernicus are among the most singular upon the moon, for they inclose a group of irregular-shaped plains, the walls of which consist of immense, more or less separate, masses. Look at the one nearest to Copernicus, which has somewhat the form of a starfish, and observe how curiously its southern border reflects, on a smaller scale, the forms characteristic of the headlands and bays along the shore of the *Mare Imbrium* below.

"Above Copernicus you see a large crater ring more than half in shadow, with a plain of an irregular hexagonal shape, northwest of it. The large ring is named Reinhold. A broken mass of mountains extends from its southern side far into the *Mare Nubium*. In the upper right-hand corner of the picture is another large ring called Landsberg. In the upper left-hand corner you see a roughly hexagonal ring plain, level on the interior, named Gambart. Mountains break the level of the *mare* both south and north of Gambart. Those on the north are remarkable for the darkness of the surface, especially in the northwestern part.

"Almost directly west of Copernicus lies an exceedingly singular object. It is a part of the underworld of the moon, the buried moon, which was covered up ages ago by that immense outgush of lava of which I have so often spoken. Once evidently it was a ring larger than Eratosthenes. Now, only its outlines can be traced, the whole immense depression of the interior and the surrounding walls to their very top having been covered up. It is pitted and surrounded with little craters of a later date. I have already told you that Eratosthenes, the ring at the left-hand edge of the photograph, marks the termination of the great range of the lunar Apennines. But these mountains seem to be continued beyond Eratosthenes in two short branches, one turning eastward toward the Carpathians, and the other reaching to the highest part of the buried wall of the submerged ring that we have been talking about and which bears the name of Stadius. You will be interested in knowing that southwest of Stadius, but off the edge of the picture, there is a place in which low hills and ridges abound, where the German astronomer Schröter imagined that he had discovered a lunar city! His mistake was, perhaps, natural, considering the slight power of his telescope and the strangely regular arrangement of the lines of hills which he mistook for streets."

"I regret that he was deceived."

Two Great Lunar "Seas"

The *Mare Serenitatis* and a Part of the *Mare Imbrium.*

"So do I. We shall now leave Copernicus and its marvelous surroundings, and turn to the last photograph in our series, representing the *Mare Serenitatis* in its full extent, and a large part of the *Mare Imbrium.* Is it not a beautiful picture?"

"It is, indeed, but so strange!"

"There is, I believe, nothing in the lunar world that would not seem strange to our eyes. To understand just what this picture means you should imagine yourself floating in an airship at an immense height above the surface of the moon. The *Mare Serenitatis* you will recognize as the great oval plain occupying the upper left-hand part of the photograph. It is entirely encircled by mountains except in three places—at its eastern end, where a broad strait opens between the Apennines on the south and the Caucasus on the north, leading into the *Mare Imbrium*; on the northwest, where another strait opens into the *Lacus Somniorum*, the 'Lake of the Sleepers,' or 'The Dreamers,' and on the southwest, where a third strait with a conspicuous crater in its center leads into the *Mare Tranquillitatis.* The *MareSerenitatis* is 430 miles long and nearly as broad, and covers an area of about 125,000 square miles. A great many details are visible on its floor. Even if it were covered with water we might see these, for, as you have probably heard, the bottom of deep lakes is visible when one looks down upon them from a great height. The surface of water, however, at certain angles of view and of illumination, would produce flashes and glares of light which are never seen on this vast lunar plain."

"Oh, but it *must* once have been a sea," said my friend, poring over the photograph. "I cannot give up that idea. It gives the interest of life to the moon, if not now at least in the past."

"You are by no means compelled to give up your idea," I replied. "On the contrary you are supported by the opinion of many astronomers, including Messrs. Loewy and Puiseux, whom I quoted a little while ago. They aver that the resemblances between the lunar *mares* and the beds of our terrestrial oceans are too numerous and too decided to permit any other conclusion than that in the one case as in the other a deep covering of water has produced the characteristic features. One striking resemblance that they note is in the surface contours. The lunar sea beds are generally deepest along the shores; the same is true of the terrestrial seas. Continents on the other hand are characterized by concave surfaces. But before we study the two lunar 'seas' in detail let us first look at their shores and surroundings. The upper and right-hand sides of the *Mare Serenitatis* are bordered by hundreds of miles of magnificent cliffs, which in many places are very steep and of great height. These form what we may call the sea front of the Hæmus Mountains, which join the lunar Apennines on the southern shore of the strait leading into the *Mare Imbrium*. These mountains possess one conspicuous crater, set like a gem in the chain, at about a third of its length from the western end. This crater is Menelaus, which we saw in one of the smaller photographs. It is characterized by its exceptional brilliance as well as by the fact that the longest of the bright bands that start from Tycho passes through it, and then continues on across the *Mare Serenitatis* and the *LacusSomniorum*, to the *Mare Frigoris*. This band, more than 2,000 miles long, has come all the way from Tycho, high in the southern hemisphere, never turning aside to avoid anything in its path. Mountains, craters, and ring plains are equally indifferent to it. It is like a Roman road, and like that, too, it suggests for its creation a power that knew no master, and admitted of neither rivalry nor opposition. The existence of this mysterious band increases the difficulty of finding a satisfactory explanation of the Tychonic rays. In the midst of the *mare* the band or ray crosses another lone crater, 14 miles in diameter, named Bessel. The full length of the ray is not shown in this photograph, but on its way from Bessel it touches two other small craters in the 'sea.'

"That portion of the Hæmus range in which Menelaus is set is a very attractive scene on account of the bow shape of the mountains, and the situation of the bright crater just in the center of the bow. Menelaus and the streak from Tycho can be seen at Full Moon with no greater optical aid than that of a good binocular. On the edge of the 'sea,' off a lofty headland of the Hæmus chain, another lone little crater is visible, Sulpicius Gallus by name. It, too, is remarkable for its brilliant reflective power. Behind the mountains, directly back of Sulpicius Gallus, and lying in an upraised part of the *Mare Vaporum*, is a larger, and even brighter, crater ring than Menelaus. It is named Manilius, and is likewise a conspicuous object for a binocular at Full Moon. Below Sulpicius Gallus the Hæmus Mountains broaden out and assume a curious somber tone, until, in the form of a rough plateau, they blend with the wide-expanded southwestern slopes of the Apennines. The latter rise gradually to the chain of huge peaks fronting the *Mare Imbrium.* They contain one notable crater ring named Marco Polo, which lies just above a great square *massif*, which breaks the narrow chain of the illuminated summits of the Apennines. The precipitous front of this range appears very brilliant in the afternoon sun, for here again we have a photograph made after the time of Full Moon. The end of the Apennines touching the strait, of which I have previously spoken, terminates with a high cape called Mount Hadley. In the strait, off this cape, is an array of small mountain peaks, which must have been islands, if the lunar 'seas' were once true seas.

"Across the strait, on the northern side, stand the lunar Caucasus Mountains. They run out to a point in a long, irregular, broken ridge. The distance from Mount Hadley across the strait to the projecting point of the Caucasus range is about 50 miles. The islands narrow the main opening to a width of 30 miles. In strict fact the Caucasus range is not continuous. The point fronting the strait is, in reality, the end of a large irregular 'island,' with intricate channels separating it from the mainland. Still farther north the photograph shows a broad valley severing the mountain range from side to side. The main mass of the Caucasus continues northward to the great ring mountains Eudoxus and Aristoteles. In the center of the range, opposite the lower corner of the *Mare Serenitatis*, is an irregular ring plain, Calippus. West of this the mountains break down in great precipices to the level of a plain that might be compared with one of the 'parks' of Colorado. Beyond this, in the shape of a broad mass of hills, it skirts the border of the *Mare Serenitatis* for nearly 200 miles to a sharp promontory which shuts off the *Lacus Somniorum* on one side from the *mare.* West of Aristoteles and Eudoxus the mountain mass extends to a curious sharp-angled plain, which it skirts on the north and south.

"The western shore of the *Mare Serenitatis* beyond the strait opening into the *Lacus Somniorum* is bordered by a series of alternating ring plains and connecting mountains. The first and largest of the rings is Posidonius, an immense formation 62 miles in diameter, with a central crater and curious ridges within the inclosure. Above Posidonius is Le Monnier, a ring plain whose 'seaward' wall has been broken down. Above that, again, is a mountain range terminating with broken crater rings. Then we arrive at the strait opening into the *Mare Tranquillitatis*, which is twice as broad as that between the Apennines and the Caucasus, and just in the middle of it stands a very perfect crater ring named Dawes. On the eastern side of this strait the Hæmus Mountains begin with a long cape called the Promontory Acherusia.

Above this promontory, at the edge of the picture, appears the ring plain Plinius, with a distinct central peak. This completes the circuit of the *Mare Serenitatis*.

"We return to the Caucasus region. These mountains front the *Mare Imbrium* along the upper part of their course with sharp slopes and cliffs. In the 'sea,' nearly opposite the deep, broad valley which I pointed out as dividing the range completely across, stands a triangular-shaped ring plain dark with shadow on one of its sides. This is Theætetus, interesting as the scene of an alleged display of 'smoke,' reported to have been witnessed by a French observer with his telescope a few years ago. Several occurrences of this kind have been reported on the moon, but more or less doubt attaches in every instance the accuracy of the observations, or at least to that of the conclusions drawn from them. Below Theætetus is an oval ring almost entirely filled up, with two craters within it. This is named Cassini. Below Cassini begins another mass of mountains, the lunar Alps. These are by no means as extensive as the Caucasus, but they contain some lofty peaks, and are traversed by one of the most remarkable valleys on the moon. It is not very distinctly shown in this picture, but you may recognize it by a dark band commencing opposite a small bay which sets back into the mountains. The valley continues through the mountains and the adjoining hilly regions nearly to the shore of the narrow *Mare Frigoris*, which runs in a sloping direction from beyond Aristoteles to the bottom edge of the picture. The Alps spread eastward, broadening out with many separate peaks, and skirting the *Mare Imbrium*, until they reach one of the most singular and interesting of lunar formations, the oval ring plain Plato. This looks like a dark lake surrounded by high cliffs. In the photograph all of the encircling wall is illuminated on the inner side except at the east end, where the shadows extend a short distance upon the floor. Plato looks as though it might once have been a ring mountain of the usual type, which has been partly filled in the interior by a local uprush of molten lava. The diameter of the ring is 60 miles, but the inclosure sinks only about half as deep beneath the crest of the wall, as is the rule with formations of similar outline. A central peak, a group of mountains, may be buried there.

"It is within this ring of Plato that some of the strongest evidences of continued change, and possibly of continued life upon the moon, have been found. Prof. William H. Pickering, after long and careful studies of this remarkable plain, says of it:

"'Plato is, I believe, more active [in a volcanic sense] than any area of similar size upon the earth. There seems to be no evidence of lava, but the white streaks indicate apparently something analogous to snow or clouds. There must be a certain escape of gases, presumably steam and carbonic acid, the former of which probably aids in the production of the white markings.'

"The white marks to which Professor Pickering refers are but faintly indicated in the photograph before us, but with the telescope, when the illumination is favorable, they are plainly seen. There are a number of very small crater pits scattered over the floor of Plato, and around these changes of color occur which have been ascribed to the emission of some substance from the pits and to the presence of vegetation, nourished by the gases and vapors, and springing into renewed life every time the sun rises upon the plain. Broad areas of the inclosure gradually change color as the sun rises, and again as the sun sets, and these phenomena have also been ascribed to the presence of vegetation. You may, if you wish, regard Plato as a kind of mountain-ringed prairie, covered with something analogous to prairie grass and shrubs, which depends for its existence, partly, upon the supply of gases spreading over the surface from the crater pits."

"So this, then, is your 'lunar grass'?"

"Yes, but not all of it. Mark, I do not aver that it actually exists; I only say that it has been suspected to exist. On some of the *mares* similar appearances are seen, as I have already told you, on a much more extensive scale, and I may again quote Professor Pickering, who says that some of his observations 'point very strongly to the existence of vegetation upon the surface of the moon in large quantities at the present time.'"

"Does this vegetation resemble that of the earth?"

"I cannot tell you."

"But where vegetation exists animal life is possible, is it not?"

"Yes, it is *possible*."

"What forms would it have?"

"I cannot tell you. But I certainly should not expect to find manlike creatures there."

"Oh, men are not *necessary* everywhere," said my friend, laughing. "I am content if you admit that there may be living creatures of some kind. Henceforth I shall never forget Plato and the other places on the moon where such significant changes are seen."

"I shall presently point out to you one of the most notable of those other places," I replied. "Let me now fulfill my promise to tell you more about the lunar atmosphere. I have told you already that there are strong reasons for supposing that the moon once had a far more dense atmosphere than she possesses at present, and I have mentioned some of the ways in which this atmosphere is supposed to have disappeared. I think that it is worth our while to refer to them again. In the first place the moon's atmosphere may have been withdrawn into vast internal cavities formed by the gigantic volcanic eruptions. Secondly, it may have been absorbed both mechanically and chemically by the core of the moon as it cooled off. We know that cooling rocks absorb immense quantities of the gases constituting the air we breathe. In fact we may look forward to a time, fortunately for us extremely remote, when the interior rocks of the earth will, in this manner, absorb perhaps all of its atmosphere."

"But if the air of the moon has gone into great cavities in the interior, why might not the living beings of the moon have followed it there?"

"According to some of the theorists," I answered, "that may really be what has occurred, and thus the moon has become a 'cavern world' on a gigantic scale. But science does not regard seriously these speculations about 'cave life' in the moon. A third hypothesis is that which I have mentioned concerning the escape of the atmospheric gases from the moon on account of its attraction being insufficient permanently to retain them. This process would be gradual, because the molecules of a gas fly in *all* directions, only a small proportion having their trajectories directly away from the center of the globe on which they are held. But a singular consequence of this theory is that interplanetary space must contain an enormous number of such wandering molecules, and every attracting body must draw more or less of them to its surface, thus forming an atmosphere for itself. As Professor Young has remarked, if as many of these molecules enter a planet's atmosphere in a day as escape from it there can be no decrease of the total amount of air. If more escape than enter, the atmosphere will diminish. If more enter than escape, the atmosphere will grow. Finally if none escape the atmosphere may increase indefinitely. This, as far as the effect of gravitation is concerned, should be the case on the sun, for the solar attraction is more than sufficient to retain any gas known to us. In consequence, the sun's atmosphere may be increasing in extent and density. Even the earth's atmosphere may be slowly increasing from this cause, and herein may lie the explanation of the enormous atmosphere surrounding the great planet Jupiter.

"In view of what I have said it is evident that the moon cannot be entirely airless. Recent observations have confirmed this conclusion, and some observers have thought that they could detect the presence of something resembling clouds occasionally creeping like low fogs over certain places on the moon. All this, you will observe, has an important bearing upon the question of life on the moon at the present day. Certain forms of plant life and low animal organizations might exist in such an atmosphere as the moon still possesses."

"But," interjected my friend, "is not this that you have been telling me in contradiction to what you said about the cause of the sharp division between day and night on the moon, and about the visibility of the stars there in the daytime?"

"Not at all," I replied, "for the effects of which I spoke are relative. In any case the atmosphere of the moon must be too rare to diffuse any perceptible amount of light into the shadows, or to illuminate the sky sufficiently to render the stars invisible. The same reasoning applies to what I have told you about the contrasts of cold and heat on the moon.

"But we have not yet finished with our photograph. We were looking at the plain of Plato, you will recollect. Notice, now, the *Mare Imbrium* off the coast that adjoins Plato on the south. You see there several bright spots resembling islands. Islands they must have been if the *mare* once had water covering it. One of these, standing by itself, an irregular, bright clump with a distinct shadow on the western side, bears the name of Pico, taken from the sharp peak in the Azores Islands. The broken mass southeast of Pico, and nearer the coast, constitutes the Teneriffe Mountains. You will notice that terrestrial geography has been drawn upon in this case also to supply a name. Still farther east is a long 'island' named the Straight Range. Beyond that, at the edge of the picture, appears Cape Laplace, at the western end of the 'Bay of Rainbows.'

"We now turn to the southwestern border of the *Mare Imbrium*, in the upper part of the photograph. This, as I have already pointed out, is skirted by the steep cliffs of the Apennines for a distance of more than 400 miles. Opposite the crater ring Marco Polo, in the Apennines, you will notice how the floor of the 'sea' is upheaved, containing a great number of irregularities, and some small peaks. This would have been a dangerous part of the 'Sea of Rains' for the lunar navigators. At the northwestern corner of this region lies a large ring plain, with indefinite light stripes crossing its floor, which is named Archimedes. It is about 50 miles in diameter. Northwest of it are two smaller ring mountains, Aristillus (the larger) and Autolycus. If we could suppose these immense volcanoes to have been in eruption when these seas were navigable, imagine the magnificent spectacle that they would have presented to anyone approaching in a ship from the direction of the strait between the Apennines and the Caucasus.

"Let us now pass this strait and enter the *Mare Serenitatis*. You will admire the beautiful modulation of the bottom, as shown in the photograph. Lighter and darker regions are curiously interspersed, and in some places there are faint indications of that wonderful lunar world of remote antiquity which lies buried in the grave of a planet. Directly opposite the opening of the strait, a small, round, light spot is seen in the midst of the sea. This is Linné, very famous for its strange and suggestive history. Here, if anywhere on the moon, changes visible to human eyes have taken place, and, in the opinion of Professor Pickering, are still taking place every fortnight. In the center of the light spot is a minute crater, and from this crater there seems to issue some kind of vapor which spreads over the surrounding surface, alternately expanding and shrinking in extent. A remarkable change in the form and appearance of Linné was recorded by the astronomer Schmidt, at Athens, in 1866. What had occurred has been explained by some as the falling in of a crater floor some six miles in diameter. But the observations of Professor Pickering are more interesting and suggestive. According to him the bright patch about the crater pit extends during the lunar night and diminishes by day, indicating that something issues from the pit and is deposited over the surrounding plain in the form of hoar frost, which melts away in the sunshine. He has even recorded an apparent expansion of the white area during a lunar eclipse when the cold shadow of the earth tends to condense the vapors. If this is true it seems rather surprising that many more similar phenomena are not visible elsewhere.

"Among the most remarkable and beautiful features of this photograph are the winding ridges like half-submerged mountain ranges that appear on the sea bottom in various places. Notice particularly the long twisted chain that lies across the western part. Between this and a shorter range, close to the west shore, runs a broad, dark valley, with the crater Dawes lying in the middle of it at the upper end. Some of these winding ridges suggest by their shape and modulation the action of water. Finally, let us return to the strait through which we recently passed. Notice that the Apennines and the Caucasus look as though they had once formed a continuous line of mountains, which has been broken through in its center, leaving huge buttresses on each side, like the Pillars of Hercules at the Strait of Gibraltar?"

"That place has an irresistible attraction for me," said my companion. "I cannot withhold my imagination from picturing the scene there when the waters rolled deep over those great bottoms, and when white-sailed ships were passing and repassing between the towering capes, carrying the commerce of opulent cities situated along those marvelously picturesque shores."

"Perhaps," I suggested, "the lunarians, whom you have reconstructed in your fancy, reached, before the catastrophe came that ended their existence, a higher state of civilization than ours, and learned to substitute electrically driven vessels for white-winged ships."

"That would be like the introduction of vulgar steamboats on the canals of Venice," she replied.

"Well," I said, "this ends our survey and one month of photographic journeying on the moon, and I am glad that you have finished it with so pleasing a vision."

Upon parting from my friend I left the photographs in her possession. A few weeks later I received a letter from her in which she said:

"I have been studying and restudying those wonderful pictures of the moon. I have ordered a telescope to be set up in my park near the elm, and when it is ready I wish you to come and instruct me how to view the moon for myself. I believe that I am becoming a learned and enthusiastic selenographer, and those strange names—Gemma Frisius, Bullialdus, Abulfeda, Abenezra, Rabbi Levi, Maurolycus, Fra Mauro, Sacrobosco, Zagut, Cichus, Sulpicius Gallus—have established their fascination over my mind. Theophilus no longer terrifies me with its formidable aspect, and I spend hours poring over the *Mare Serenitatis*. But my fancy remains faithful to the 'Marsh of a Dream,' which still represents for me the culmination of lunar ideality.

"As to life on the moon, I find that I cannot be satisfied with a mere grass theory. I am so well convinced that there must be something more, that I no longer relegate my lunarians to an age antedating the volcanoes. On the contrary, as soon as I get my telescope I am going to look for signs of them and their doings in the present day, and willy nilly, sir, you have got to aid

me in the search."
    APPENDIX

APPENDIX

Dates, and age of the moon, when the twenty-one serial photographs were made at the Yerkes Observatory, by Mr. Wallace, with the 12-inch telescope and a special color filter constructed by him:

No. 1 , - February 19, 1904; - Moon's Age 3.85 Days

No. 2 , - September 24, 1903; - Moon's Age 3.87 Days

No. 3 , - July 29, 1903; - Moon's Age 5.54 Days

No. 4 , - November 24, 1903; - Moon's Age 5.74 Days

No. 5 , - July 1, 1903; - Moon's Age 6.24 Days

No. 6 , - November 26, 1903; - Moon's Age 7.75 Days

No. 7 , - July 2, 1903; - Moon's Age 7.24 Days

No. 8 , - August 31, 1903; - Moon's Age 9.22 Days

No. 9 , - August 2, 1903; - Moon's Age 8.97 Days

No. 10 , - November 30, 1903; - Moon's Age 11.78 Days

No. 11 , - December 1, 1903; - Moon's Age 12.98 Days

No. 12 , - September 4, 1903; - Moon's Age 13.27 Days

No. 13 , - September 5, 1903; - Moon's Age 14.40 Days

No. 14 , - August 26, 1904; - Moon's Age 15.65 Days

No. 15 , - August 28, 1904; - Moon's Age 17.41 Days

No. 16 , - August 29, 1904; - Moon's Age 18.62 Days

No. 17 , - October 10, 1903; - Moon's Age 20.06 Days

No. 18 , - September 29, 1904; - Moon's Age 20.50 Days

No. 19 , - August 16, 1903; - Moon's Age 23.81 Days

No. 20 , - August 17, 1903; - Moon's Age 24.84 Days

No. 21 , - August 19, 1903; - Moon's Age 26.89 Days

INDEX

INDEX

BOOKS BY GARRETT P. SERVISS.
Other Worlds.

Their Nature, Possibilities, and Habitability in the Light of the Latest Discoveries. Illustrated. 12mo. Cloth, $1.20 net; postage, 11 cents additional.

This book presents the very latest conclusions in regard to the nature and the habitability of the other planets. It is written in popular style, and, at the same time, is scientifically accurate in its statements. It is a convenient handbook of information concerning the solar system, but by no means a dry, scientific treatise on the subject. It might be said to resemble Proctor's celebrated "Other Worlds than Ours" brought up to date. The last chapter, on "How to Find the Planets," is unique and should prove very useful.

Pleasures of the Telescope.

A Descriptive Guide for Amateur Astronomers and all Lovers of the Stars. Illustrated with charts of the heavens and with drawings of the planets and charts of the moon. 8vo. Cloth, $1.50.

"This is a book which will give intense pleasure to everyone who uses it and follows its clear instructions."—*Louisville Courier-Journal.*

"Every person of culture should possess at least a passing acquaintance with the planets, stars, and constellations. With a little patience and comparatively small effort Mr. Serviss's new book will enable anyone to obtain this knowledge."—*Los Angeles Herald.*

Astronomy with an Opera-Glass.

A Popular Introduction to the Study of the Starry Heavens with the Simplest of Optical Instruments. 8vo. Cloth, $1.50.

"We are glad to welcome this popular introduction to the study of the heavens.... There could hardly be a more pleasant road to astronomical knowledge than it affords.... A child may understand the text, which reads more like a collection of anecdotes than anything else, but this does not mar its scientific value."—*Nature.*

D. APPLETON AND COMPANY, NEW YORK.

BOOKS ON ASTRONOMY.

Popular Astronomy.

A General Description of the Heavens. By Camille Flammarion. Translated from the French by J. Ellard Gore, F.R.A.S. New Revised Edition. With 3 Plates and 288 Illustrations. 8vo. Cloth, $4.50.

The author is the most popular scientific writer in France, and the present work was considered of such merit that the Montyon Prize of the French Academy was awarded to it. The subject is treated in a very popular style, and the work is at the same time interesting and reliable. The work has been newly revised throughout and an appendix added, showing all the important advances made in astronomy up to the year 1907.

The Earth's Beginning.

By Sir Robert Stawell Ball, LL.D., F.R.S., author of "The Story of the Sun," "An Atlas of Astronomy," "Star-Land," etc.; Lowndean Professor of Astronomy and Geometry in the University of Cambridge; Director of the University Observatory, etc. With four colored Plates and numerous Illustrations. 12mo. Cloth, $1.80 net; postage, 18 cents additional.

This book will make admirable reading for persons of any age. It provides a clear and popular explanation of the great problem of the earth's beginning. It is believed that no exposition of the nebular theory and its infinitely wide ramifications has been made that is at once so simple, so original, and so comprehensive. Dr. Ball's success as a lecturer in this country indicates his gift of popular exposition, and this book will rank as one of the most attractive presentations of scientific fact and theory for general readers.

The Sun.

By C. A. Young, Ph.D., LL.D., Professor of Astronomy in Princeton University. New and revised edition, with numerous Illustrations. 12mo. Cloth, $2.00.

The Expanse of Heaven.

By Richard A. Proctor. A Series of Essays on the Wonders of the Firmament. 12mo. Cloth, $2.00.

Other Worlds Than Ours.

By Richard A. Proctor. The Plurality of Worlds, Studied under the Light of Recent Scientific Researches. With Illustrations, some colored. 12mo. Cloth, $1.25.

Light Science for Leisure Hours.

By Richard A. Proctor. A Series of Familiar Essays on Scientific Subjects, Natural Phenomena, etc. 12mo. Cloth, $1.75.

The Story of the Stars.

By G. F. Chambers, F.R.A.S., author of "Handbook of Descriptive and Practical Astronomy," etc. With 24 Illustrations. (Library of Useful Stories.) 16mo. Cloth, 35 cents net; postage, 4 cents additional.

D. APPLETON AND COMPANY, NEW YORK.